Construction Companion to Risk and Value Management

Peter Walker and David Greenwood

RIBA Enterprises

© Peter Walker and David Greenwood 2002

Published by RIBA Enterprises Ltd, 1–3 Dufferin Street, London EC1Y 8NA

ISBN 1 85946 092 5

Product Code: 23443

British Library Cataloguing in Publication Data
A catalogue record of this book is available from the British Library.

Publisher: Mark Lane
Editor: Lionel Browne
Series Editor: David Chappell
Commissioning Editor: Matthew Thompson
Project Editor: Katy Banyard
Design: Bettina Hovgaard-Petersen

Typeset, printed and bound by Hobbs the Printers, Hampshire

Contents

Foreword

This is the fifth in the *Construction Companion* series published by RIBA Enterprises. These books are intended to be compact and accessible guides written in plain language but each one authoritative in its own subject.

Risk and value management is probably not the favourite topic for most architects. It is the kind of thing that induces either deep apathy or flight. Yet an understanding of the principles is absolutely essential to every phase of the building process. Whether they know it or not, construction professionals spend most of their time first identifying and then trying to manage risk and value.

Although many would say that they understand the meaning of risk (and they probably do, even if their understanding is somewhat restricted), most people do not understand the meaning of value at all. There is scarcely a decision that the architect is called upon to make that does not involve allocating and balancing risk, looking at the value of things, and devising methods of adding to that value.

A question often asked is, 'Do building contracts have to be fair?' The answer, of course, is no; but they must allocate risk between the parties in a way that both parties – probably for entirely different reasons – find acceptable. Much of the opportunity for value management occurs in the early stages of a project. Adding risk may add value in certain circumstances.

Although it is vitally important to understand these concepts thoroughly so that they can be employed to best effect, architects are not attracted by the kind of jargon that seems to be a feature of most textbooks. This book is written for those people. Peter Walker and David Greenwood have done an excellent job of explaining risk and value management in a relatively simple way. That is not to say that the book is devoid of unfamiliar words, but new words and concepts are explained as the book progresses.

This is an impressive book for anyone needing to learn more about the subject, and that should be all construction professionals.

David Chappell BA(Hons Arch) MA(Arch) MA(Law) PhD RIBA
Series Editor

David Chappell has worked as an architect in the public and private sectors and is currently Director of Chappell-Marsh Limited, a building and civil engineering contracts consultancy, and Professor of Architectural Practice and Management Research at the Queen's University of Belfast. He is a frequent conference speaker and author of many books for the construction industry.

Preface

This book is about managing risk and value in construction projects. But, more basically, it is about decision-making. In a recent workshop on the subjects of risk and value,[1] Professor Patrick Godfrey of Halcrow Business Solutions asked the following questions. How can an organisation, he mused,

- properly define a project without understanding its purpose and objectives?
- reliably decide what it is going to do about it without considering the risks?
- effectively assess the risks without knowing the objectives?

These three questions neatly close the logic loop between the two headline topics in this book: risk and value. Our approach to the two is that they are both indispensable and inseparable.

Chapter 1 of the book is an introductory chapter that examines where construction risks come from, and what the concept of value means. This is followed, in Chapters 2 and 3, by an exploration of the process of managing value and risk in two different stages in the life of the project. Chapter 3 deals with the design function, from inception to the start of construction, and Chapter 4 deals with the largely on-site function of constructing the project. Chapters 5 and 6 examine the techniques and procedures involved in using value management and risk management as tools for construction project management.

Chapter 7 deals with a very topical issue: whole life or life cycle costing. Previously largely neglected, this aspect of value will become hugely significant as we enter an era in which few construction developments will be able to proceed without first paying attention to matters of sustainability.

Chapter 8 is a brief digression into the social and psychological aspects of risk and value. It is important to remember that neither individual nor group decision-makers always behave predictably, particularly in conditions of uncertainty.

It is probably fitting that the final chapter of the book, Chapter 9, covers a trend that may dominate the future of risk and value management: the significant trend away from piecemeal tendering and towards partnered approaches to procurement. This includes coverage of both project-specific and serial partnering, and the so-called public–private partnerships epitomised by the Private Finance Initiative.

Notes and references

1 *Integration of Risk and Value: The Route to Best Value* CPN Members' Report 1127 (Construction Productivity Network, June 2001).

1 Risk and value in the construction process

1.1 Introduction: managing construction projects

Construction is a project-based industry. There are exceptions, of course, but usually the focus of attention is the project.

Projects need managing, and a lot has been written about how to do it. A common theme is that there are three principal factors to project success: time, cost and quality. Some, quite properly, add a fourth: safety. We might call these the first-line objectives of project management. They could be represented as in Figure 1.1.

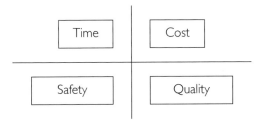

Figure 1.1 The first-line objectives of project management

So, managing a project is about managing these first-line objectives: it couldn't be simpler. Yet projects *do* go wrong; and sometimes it seems that the bigger they are, the more wrong they go. We shall look more closely at this 'big-project syndrome' later.

Perhaps the project management manuals are a little optimistic. A classic illustration of this is the so-called PRINCE project management methodology, which prescribes the process of organisation, management and control of projects. PRINCE was originally developed for managing IT projects, a risky environment if ever there was one. From its IT origins, PRINCE has spread into many industries to become the UK generic standard for project management. And yet the name itself betrays a rash degree of optimism about the difficulties of managing projects: PRINCE stands for Projects in Controlled Environments!

1.1.1 Projects as systems

Systems theory has been one of the most significant advances in the study of the way organisations work. Organisations can be viewed as a collection of interrelated functions, and to understand such a *system* properly requires

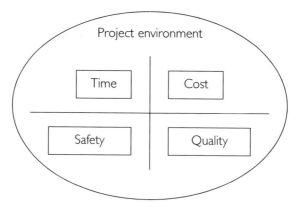

Figure 1.2 A project as a (closed) system

knowledge both of the parts *and* of the way they interreact. The projects we are trying to manage can be also viewed as systems, and Figure 1.2 shows a simple version of such a view.

But this is not the whole picture. Most observers of projects would agree that they are not 'closed' systems. It may be quicker, easier and more convenient to work that way, but ultimately some account must be taken of the external environment (the word is here used in its general, rather than ecological, sense) within which systems operate. A further contribution of *systems thinking* is that for systems to work properly they must have access to, or take note of *feedback* from, this environment. In other words systems are best thought of as *open*, not *closed*. As will be seen later, if we neglect this, and assume that the environment is in a 'steady state', the result can, for some projects, be disastrous.[1] Figure 1.3 represents

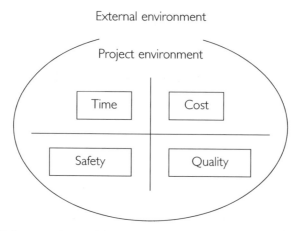

Figure 1.3 An open system model

this simple but important development from a *closed* to an *open* systems view of a project.

1.1.2 Managing the system

It is often said that the most difficult and possibly most critical area of management is *interface management*: that is, managing at the 'edges' of systems. These edges, or interfaces, could be physical: between two components in a building, for example. But the interfaces in question in the current model are those between its constituent parts: time, cost, quality, safety, and the outer environment. The management of value and the management of risk are just such forms of interface management. Take value management, for example. Whereas the principal project objectives are individually to manage time, cost, quality and safety, one could conceive of value management as taking place at the interfaces between these first-line objectives.

Example

A client requires a new building, and has heard that the quickest method of procurement is design and build. He follows this route, and because he is interested in keeping the cost down, he has an outline specification produced and invites tenders on a guaranteed maximum price basis. When the contractor begins work the client notices that the product is not quite as he imagined and that the quality is 'basic'. Should he be surprised? It is relatively easy to keep a tight rein on any two of the project objectives mentioned above (in this case time and cost) but then another (here, quality) can be compromised. To strike the required balance between all of them is not easy.

The objective of value management is to reconcile these different demands and maintain the appropriate balance. Risk management, by contrast, occurs at the interface of these demands with the wider project environment, and with the wider-still 'external' environment: the principal objective is to mitigate the effects of the unknown that lurks outside the system. A later part of this chapter will examine what these effects are.

1.2 The concept of risk in a construction project

1.2.1 Introduction

So far, we have adopted a simple approach to risk. Risk management is mitigating the effects of the unknown. There is *risk*, there is *us*, and we try to manage it. This is the scenario at its most basic. But human endeavour is rarely so simple as individuals and groups working cooperatively towards a common purpose. In the real world, projects are accomplished by a coming together of people who buy and sell their products and skills – and while they are doing this these people also buy and sell risks.

In commercial terms risk is normally allocated through the medium of contracts. Some contracts – those concerned with insurance – are exclusively concerned with allocating risk. Insurance contracts are widely used in construction. But the construction contracts themselves are different: they are about defining what is to be done and what is to be paid for doing it. Yet concerns about risks are never far away. Many of the contract conditions that attach to a construction project are actually about allocating risks. There are choices between using the norms of risk allocation created in the so-called standard contract forms; there are the effects of departing from these norms in terms of cost, effectiveness and efficiency; and there is the degree to which the state will intervene on grounds of unfairness. There is a need to recognise the incentives that are deliberately, or sometimes unknowingly, put in place by different approaches to contractual risk allocation. These issues are considered in greater detail in Chapter 2.

Theories of contractual risk allocation usually conclude (1) that risks are best allocated to the party that is most suited to managing them, and (2) that somewhere there will be a price for doing so. Intimately connected with this price is the notion of risk *attitude*. Whether we are dealing with individuals or with organisations, their attitudes to risk will differ, and it is this difference that makes them feel able to take or to shed risk for a given price. Economists would say that it is this difference in attitude that allows parties to enjoy mutual gains by shifting risks from one to another: such deals involve finding the ideal amount of risk to shift, given the parties' risk attitudes and the costs of shifting it. The issues that surround risk attitude and risk preference will be introduced in Chapter 8. Chapters 2 and 3 will focus on these same issues, but specifically from the perspective of the participants in the construction process.

1.2.2 Where do construction risks come from?

In construction, risks can come from many sources. They can arise from difficulties with the *physical design* of the product itself, perhaps from using an innovative structural technique, or a material that has never been used before. Or they may lie not in the *product* itself but in the way it is planned to construct it, or the basis for raising the money to do so. The risks may come from unspecified sources outside the project itself – what we have referred to as the project *environment*. And finally there may be *political* risks, in the form of changes in legislation, changes

Table 1.1 The sources of construction risks

Issue	Examples
Political/regulatory	Government. Planning. Contractual. Pressure groups. Health and safety. Environment
Technical and physical	Logistics. Geology. Topology. Weather. Culture and religion
Conceptual	Client organisation. Project scope and definition. Design
Financial	Funding. Tax. Financial stability. Inflation. Exchange rates. Bonds
Organisational	Estimating. Scheduling. Mobilisation. Project structure. Control. Communication.
Operational	Quality. Safety. Innovation. Design change. Labour. Materials. Plant. Subcontractors

in government priorities, or even changes in the government itself. Table 1.1 is one example of the many ways in which these risks can be classified.[2]

However we categorise the risks, the construction industry has the reputation for having plenty of them. The industry also has the following characteristics, which accentuate its inherent risks:[3]

- Projects tend to be large.
- Projects tend to be of long duration.
- Projects tend to be complex.
- There is a high level of subcontracting.
- Construction work is labour-intensive.

Examples
There are various examples of the relationship between the size of a project (in terms of its money value) and the tendency for it to run out of control. This relationship, known colloquially as the 'big-project syndrome', is vividly illustrated in Peter Hall's account of project failures and near-failures.[4] Hall cites the Sydney Opera House and the Thames Barrier in London. To these could perhaps be added a number of recent examples, including the Trans-Alaska Pipeline project, the Channel Tunnel, and the British Library.

Big projects also tend to take a long time, and this in itself is a source of risk, since both the supply and the demand sides of the industry are vulnerable to changes in market conditions. At worst this can result in a project where the participants are ruined by rampant inflation (like many of the projects undertaken around the time of the two 'oil price shocks' of the 1970s) or one that has become useless to its owner by the time it is finished. One way to overcome the threat of change is to attempt to deliver the project in an accelerated timescale – a practice known as *fast-tracking*. However, compressing normal durations, overlapping activities that are normally sequential, and limiting the periods for pre-project design and planning are all sources of considerable risk.

Subcontracting is a prevalent and important element of the UK construction process. Such is its extent that a recent report[5] pronounced that 'all construction work involves either subcontracting or separate trades contracting'. Subcontract relationships are crucial to project success but can be extremely problematic, with adversarial and often hostile relations between the parties, which often coincide with project delays, cost overruns, defective quality and problems with safety.

Although the construction industry has undergone significant modernisation, with a movement to mechanisation and off-site fabrication, it remains highly labour-intensive. And this relatively heavy reliance on labour means a more volatile mix of the factors of production.

It is not always easy to define what constitutes a complex project. Large projects are often, though not necessarily, complex, while most managers will concede that relatively small projects can be disproportionately complex. One measure of complexity is the number of junctions between the separate constituent parts: for example, between the technical inputs. A way to measure this might be to take a logic-linked programme, such as a critical path network, and measure the number of 'dependences'. Whatever its definition there appears to be little disputing the fact that project complexity increases project risk.

1.3 The concept of value in a construction project

1.3.1 Introduction

Those involved in designing and constructing buildings find that, in the course of a construction project, much time and energy is given to matters of *cost* and *price*. The architect, in designing to meet the project's financial budget; the estimator, in pricing the tender documents to arrive at the most competitive tender figure; and the quantity surveyor, in forecasting and managing the project budget, are all largely focused on cost (or money) issues and measures. The management processes, the forms of contract and the way in which construction transactions take place are all usually focused on *cost* rather than on *value*.

One of the traditional measures of achievement in the construction industry is delivery of the building within the budget. Construction is not unique in this, and the effective management of cost should of course be both a central concern of and an important measure of success for construction projects. The capital cost is of fundamental importance: if the required funds cannot be secured at the outset to finance a project it will not be feasible and will simply not happen. Testing the viability of a project is of course concerned with precisely that: testing the financial feasibility – the capital cost or the return on investment of a particular project.

Cost is however only one measure of value, and in many ways it is a crude measure. Value goes beyond cost, and includes matters other than simple monetary and price measures. The management of value therefore requires some articulation of and agreement with the concept of value, and this is where life gets complicated. 'Value' is one of those slippery words that crop up in many different contexts, often with distinctly different meanings. It means different things to different people, and is generally used colloquially in a much looser way than for instance an economist, surveyor or accountant would use the word.

To understand value management in construction it is necessary first to get to grips with some of the common meanings and concepts of value, and to look at the organisation of the construction industry and construction projects. Finally, the two need to be put together in order to examine value in the specific context of a construction project.

1.3.2 The idea of value

Fundamentally, value is the *worth* or *desirability* of something. An economist would go further and say that value is the *amount for which something can be exchanged in an open market*: this implies the influence of supply and demand as a determinant of value (or of cost). But if value is a measure of worth then why for example is water (which is essential for life) cheaper, measure for measure, than beer (which we can get by without)?

This is explained by the concept of *marginal utility*. The value is set not by choosing between the effects of forgoing all water or all beer (as this is an unlikely choice to ever have to make), but by the effects of forgoing an additional unit of either. So, if people were asked 'Which would you give up – all water or all beer?', the answer would be obvious. However, if people were asked to choose between giving up 150 litres of water or 4 litres of beer a month, their answers would vary. Some who don't care for beer would happily forgo the beer rather than the water; others might choose to compromise and partly cut both their water and beer consumption. Others again might decide to keep all the beer on the basis that the amount of water they were left with after giving up the 150 litres was enough to get by on if they took one less bath a month. Their decision would however always be influenced by the general supply of water: that is, whether water was plentiful (low marginal utility) or scarce (high marginal utility) at the time. Value is therefore measured relative to the wider supply of goods and services.

We also talk of *added value,* which is a measure of the change in value between the inputs to a process (labour, materials, etc.) and the outputs (the finished product). The aim is to maximise the added value: in construction, for example, the cost of the bricks and mortar, the architect's time, the land etc. should in total be less than the 'value' of the finished building. Note, however, that value is not determined by the amount of input; to put it another way, value is not created by the effort involved in making something.

Value for money best illustrates the distinction between cost and value. The cheapest solution is not necessarily the one that gives the best value for money. Paying more for something often means that it will perform better, or last longer, or require less maintenance, or at best all of these. Experience shows that this is by no means always true. A particular example of this is the brand value of a product, where the value of the brand name leads to premium pricing without necessarily any increase in value for money. Value for money is an important consideration when looking at the life cycle costs of a building, and this is dealt with more fully in Chapter 7. Note also that value for money can, and frequently does, have both qualitative and quantitative measures.

In the late 1990s the UK government introduced the concept of *best value* as a measure to be used in central and local government procurement. Guidance notes on the subject stress that best value does not mean 'lowest price', but rather the purchasing of services or goods (or buildings) that deliver the best value for money. For buildings this can be measured by reference to such matters as the long-term performance, maintenance costs, environmental and sustainability issues or productivity outputs – many of which include both qualitative and quantitative measures.

Finally, we also talk of *personal values* (sometimes referred to as *belief systems* with reference to a professional group, for example architects) and of making *value judgements.* In the context of value management these are important in two respects. First, value legitimately has both personal and subjective components that require – and this is important in management terms – qualitative rather than quantitative measures. Second, those involved in carrying out any process – in construction, for example, the builder, the architect, the client – will have both personal values and professional belief systems that inform and influence their views and actions.

Architects in particular often carry much cultural baggage that may strongly influence their views of value in a construction project. This is neither a good nor a bad thing, but it will have an impact and should be considered in the management of value in construction. For example, an architect may place a high value on the appearance of the building based on the quality of materials, the detailing of the assembly of components and the quality of the spaces created. A project manager may place a high personal value on delivering the building on time or within the original budget. In both cases the approval of their peers would be an important measure of success. This approval would reinforce the professional's belief in the importance of these matters. Viewed objectively the importance of these measures of value would vary for the client (who may place great importance on the cost of the building), the user (who may appreciate the high-quality internal finishes), and society at large (who may welcome an exciting, attractive new building in their city).

So value has many meanings and facets, all of which need to be considered in the value management process. Value cannot simply be measured by that convenient, universal quantitative device called money.

1.3.3 The nature of the construction industry and construction projects

Like the concept of value, the construction industry is difficult to define, and has many unique and unusual characteristics. To understand the concept of value in construction it is necessary to understand something of the context and the processes involved in making buildings. An in-depth analysis of the construction industry would take far more space than is available here; in any case it has been well covered by others.[6] However, there are three fundamental aspects of the organisation of the construction industry that we must briefly consider, as they have important implications for the management of value.

A key defining characteristic of the construction industry, and of the way in which projects are carried out, is the high degree of *fragmentation*. This has two aspects: the split between design and construction, and the extensive use of subcontracting in the building process. Although design and build procurement unites the design and build functions contractually, they are rarely united organisationally in the sense of the designing and building functions taking place in the same firm. Although this is a good thing in some respects — it allows greater flexibility and enables risk to be spread — it also has disadvantages. One of the most significant disadvantages is the problem of inefficient communication — electronic, written and oral — between organisations. Incompatibility of software and operating systems generally means that it is easier to e-mail or place on the network a drawing in a format that can be opened and worked on within a single organisation than it is between separate companies. It is also of course much easier to lean across and speak to the person next to you! Much of the impetus for current industry-wide initiatives such as partnering and supply chain management derives from a desire to overcome the inefficiencies (or put another way, the barriers to adding value) that arise from this fragmentation.

Compared with other industries the take up of e-commerce in the construction industry has generally been slow. As a result, opportunities to add value arising from the improved transaction environment and the more efficient management processes available from e-commerce are being missed at present. For example, although it is now possible to bid for contracts or order and pay for building materials on-line, the take-up is low. Generally speaking anything that increases the speed of transactions, or multiplies the number of buyers and sellers in a marketplace, will reduce prices and – as a result – improve value.

Another key characteristic is that construction is a *project-based industry* (which could of course also be considered as fragmentation in time). This introduces problems of discontinuity: this has implications for the efficiency of output, which in turn reduces total value at the wider industry level, but more significantly in the context of value management at a project level it means that there is little continuity of participants. Construction projects are generally carried out by temporary project coalitions – a group of companies and individuals, assembled to carry out a particular contract, and then disbanded on completion. This has clear implications in terms of communications, team working, mutual understanding and long-term shared learning. All of these have an impact both on the ability to add value and on the way value is managed in construction projects.

Finally, the construction industry is a *service industry*. Construction projects require promoters: clients, or – to use the building contract term – employers. These promoters are motivated to build for differing reasons, which will inform and influence their view on and priorities for value.

1.3.4 Value in a construction project

People put up buildings for a variety of reasons:[7]

- for consumption – in economic terms we consume buildings by living in them (i.e. housing)
- for direct financial gain – where a property developer looks to turn a profit from the income arising from developing land or redeveloping an existing building

- for indirect financial gain – where a building is used as a place to manufacture a product or deliver a service from which a profit is made (for example an office or a factory)
- for social purposes – where the building is a place to deliver a social function (for example an art gallery or a hospital).

It doesn't take much thought to realise that the promoter will in many cases have different priorities and preferences with regard to what they value, or perceive as adding value, in their buildings.

Example: adding value by design
A company manufacturing computers derives its profit from selling the finished computers. All other things being equal, the more computers that can be made for the least money the more profit the company will make. All sorts of things will contribute to the efficiency with which this is done: the skill of the factory operatives, the efficiency of the plant, the management systems, the control of product quality, and of course – most significantly from our perspective – the building. The building will contribute in two ways: functionally and technically.

The functional contribution describes how well the building functions in support of the primary production process: how well it supports or allows the manufacturer to take in, store and move raw materials about the factory and to carry out the assembly and distribution process. The technical contribution covers such things as the lack of redundancy in the building envelope. Is the building the right size? Does it have the right number of doors? Is it the right height? Too low and it won't accommodate production plant; too high and it will waste building capital cost and cost more to heat. The technical contribution also covers planning issues such as the organisation of the spaces and their sizing. Are the optimum number of roller shutter doors included to allow goods in and out? If there are too few, production will be delayed; if there are too many, the resulting capital and maintenance costs will be wasted. A building that is thoughtfully planned will allow greater efficiency in the manufacturing process than one that is not. Inefficiency – lack of value management – in the building's design will reduce output and increase overheads, both of which directly reduce the company's profitability.

If a property developer is building for direct financial gain, their primary objective will be to erect a building that will sell or let quickly and on good terms. The incorporation of features that don't contribute to this – that don't add value, and are therefore redundant – adds to the overall cost and therefore reduces profit. This includes features that are of no value to the end purchaser or tenant and therefore do not add to the desirability or marketability of the building. For example, a building designed with a large number of cellular offices when tenants require open-plan spaces will not add value. Leaving things out (to reduce cost) can equally be a problem: for example not including access floors for cabling in an office building. This illustrates how value differs from and goes beyond simple capital cost.

An art gallery with poor lighting will fail to display the paintings properly, which will defeat the *raison d'être* for the building. Similarly, poor air quality will lead to long-term damage to artworks. In this case savings in both capital and running cost will actually detract from rather than add to the building's value. These are what would quite rightly be described as *false economies*.

Example: the value of coming home to an open fire

Most houses built today incorporate some form of space heating, which means that in functional terms they do not 'need' an open fire. However, the experience of private house builders shows that an open fire adds to the marketability of the house, which is an important value to the house builder. The open fire also adds to the pleasure and enjoyment that the homeowner derives from the property, another important qualitative value consideration. At resale the homeowner will derive the same marketability benefit as the house builder. It is the enjoyment of the fire, or the contribution it makes to marketability, that gives it its value, not its technical performance, although of course that may contribute as well. An open fire is unlikely to be the cheapest or even the most efficient way to heat a room, either in running or in capital cost.

This example illustrates both the difficulties of defining and measuring value and the fact that value is often a personal and subjective matter.

There is a further complication in considering value in construction, which relates to the number and the remoteness of the stakeholders – both those with a stake in the finished building and those with a stake in the construction process.

The building's stakeholders can be thought of in two categories – primary and secondary. The *primary stakeholders* for a new hospital, for example, would be those who pay for the building (the health trust), those who manage and maintain the building (the health trust estates department), those who use the building (the patients and their visitors), and those who work there (doctors, nurses and administrators). As buildings don't exist in splendid isolation, *secondary stakeholders* would include neighbours, those who walk past the building, the local planning authority, and even those yet to be born!

There are also the *temporary stakeholders,* those involved in creating the building – designers, suppliers, builders and subcontractors. Their concerns are likely to relate to such diverse issues as buildability, safety, peer approval and quality control.

It is difficult to reconcile the value priorities of the various primary stakeholders, even before the aspirations of the secondary and temporary stakeholders are overlaid. But that is the nature of architecture, and that is what architects do; it is the reconciling of the primary stakeholders' needs and aspirations with those of wider society that makes creating buildings so difficult and of course so exciting.

1.4   The link between risk and value

It should by now be clear that there is a link between risk and value. In the context of projects, value and risk management are concerned with handling different types of aim: value is about choosing objectives, and risk is about retaining those objectives in the face of factors that threaten their achievement. However, the two aims are so closely interlinked that integrating the approach to the two can bring advantages. An integrated value and risk management approach is

- more efficient, requiring fewer interventions in the project process than if reviews are carried out separately
- more effective: it will have a greater chance of being systematic and consistent, and, following systems thinking, will benefit from considering both aspects together.

1.5   Conclusions

In this introductory chapter we have sought to integrate the concepts of risk and value by explaining the management of each, in terms of

- balancing the project's front-line objectives
- recognising and managing the project's external environment.

We have examined the concepts of risk and value at a general level and have briefly outlined some of the characteristics of the construction industry and of construction projects. We have looked at the factors that motivate people to build and in particular the way in which this may inform and influence views and priorities with respect to value. We have also touched on the difficulties that arise from project size and complexity. One aspect of this complexity is the large number of people and organisations that either have a stake in the project under construction or are affected by the building as a finished product.

In some cases the interests and the values of these stakeholders will be common; in other cases they will be incompatible. So it is necessary to capture these views and give them appropriate weighting and consideration in the briefing, design and construction process. But what can value management add to this process? Indeed does value management itself add value? Furthermore, by allowing our model to be open to the environment we have admitted the prospect of uncertainty. As a result, can value management ever be realistically addressed without consideration of risk? And how should this risk itself be managed? In the following chapters we shall look for answers to these questions.

The following are the key points covered in this chapter:

- Managing projects, at its most basic, means managing time, cost, quality and safety.
- Promoters of projects have many notions of value, and these sometimes conflict.
- They need to be identified, prioritised, and weighted.
- Managing projects also involves managing the risk of things going wrong.
- In construction there are many sources of risk, particularly in the project's external environment.
- Construction projects are particularly prone to risk because they are large, complex and lengthy.
- The concepts of value and risk are intimately related, and their management should be integrated.

Notes and references

1 As the Nobel Prize-winning economist Kenneth Arrow observed, 'Vast ills have followed a belief in certainty'.

2 Adapted from the approach taken in P.A. Thompson and J.G. Perry Engineering Construction Risks (Thomas Telford, London, 1992).

3 As noted by the economist Patricia Hillebrandt. See her two books: P.M. Hillebrandt Analysis of the British Construction Industry (Macmillan, Basingstoke, 1984) and P.M. Hillebrandt Economic Theory and the Construction Industry, 3rd edn (Macmillan, Basingstoke, 2000).

4 See P. Hall Great Planning Disasters (Weidenfeld & Nicolson, London, 1980).

5 W.P. Hughes, C. Gray and J. Murdoch Specialist Trade Contracting: A Review, CIRIA Special Publication 138 (Construction Industry Research and Information Association, London, 1997).

6 C. Powell The British Building Industry Since 1800: An Economic History (E & F N Spon, London, 1996).

7 S.D. Lavender Economics for Builders and Surveyors: Principles and Applications of Economic Decision-Making in Developing the Built Environment (Longman, Harlow, 1990).

2 Risk and value in project procurement

2.1 Introduction

To understand how value and risk management fit within a construction project it is necessary to have some understanding both of the participants and of the processes involved. In this chapter we examine how the stage is set for the construction project, in terms of:

- the participants in the process
- the way their services are procured.

In Chapters 3 and 4 we look more closely at the implications this has for risk and value.

2.2 Procurement: relationships between participants

Once upon a time one could say that buildings were *designed* by architects, *built* by contractors (with the occasional input of specialist subcontractors) and *paid* for by clients. It is no longer possible to generalise so confidently about how things are done.

There are two main reasons for this, one technical and the other economic. First, the technology of the modern building is so complex that no single firm – designer, constructor or specialist provider – can provide the spectrum of skills necessary for its production. Second, the fragmentation of the construction process, with its origins in technical diversity, has been exacerbated by economic forces. These forces have, for example, encouraged contractors to subcontract (rather than employ labour), while different economic forces have encouraged clients (particularly those in the public sector) to adopt unusual approaches in seeking funding of their projects. Within this new order, risk and value are key issues, as they, together with technical and economic trends, have forged new sets of relationships between project participants.

2.2.1 Procurement

A project's procurement system is important. It will have an impact on the organisational, financial and administrative structure of the project, and could contribute crucially to its success or failure. It has important implications for the way project risks can be allocated amongst the participants. The *management* and allocation of risk and the strategies for getting the required value in terms of the major project variables (such as time and money) are major issues in procurement.

2.2.2 Defining the project inputs

In construction the procurement arrangements for a project must bind together a collection of organisations that bring a range of different inputs to the process.

Project initiation
The organisation that initiates the project is referred to by a number of names, for example the client, the employer, the authority (for public bodies), the promoter, the owner or the purchaser.

Project finance
The simplest approach for the input of project finance is for it to be procured by the client (or project initiator). In a more radical approach the finance comes from the supply side, not the demand side. This concept is not new: the infrastructure of several countries was created in this way. In France, for example, the first concession for water supply was let to the Perrier Brothers in 1782 but taken back after the revolution. In the private sector in the UK, builders ranging from large package dealers and builder-developers to small domestic builders commonly offer finance. In recent years there have been dramatic changes in the way many public sector projects are financed. There has been a move away from direct capital expenditure on owned facilities to arm's-length outsourcing. The most familiar example of this is the Private Finance Initiative (PFI).

Design and management
The complexity of construction projects makes design a critical element. This has led some commentators to classify procurement systems as separated, integrated or management based. The separated system – in which design is procured independently of construction – is also referred to as the traditional system, or traditional general contracting (TGC). Integrated systems – in which the design and construction form some sort of integrated package – are exemplified by design and build and its variants. Management-based procurement systems, such as management contracting (MC) and construction management (CM), treat not only design but also management as a separable input to the construction process.

Construction and other inputs
In some cases these organisations may be contracted to provide other inputs. These could include responsibility for 'ownership' activities such as commissioning, operation, and maintenance.

2.2.3 The strategies available

Because of the structure of the supply side of the industry these project inputs can come from a multiplicity of sources in many possible combinations (Figure 2.1). The chosen arrangement has become known as the *procurement strategy,* and it has a fundamental influence on the construction contract. Strategic forces, including risk and value strategies, tend to push one or another option to the fore, and even cause other systems to emerge or re-emerge.

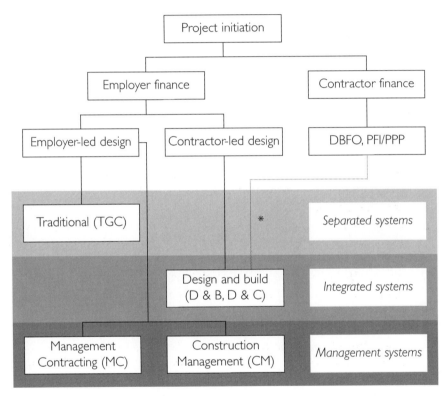

Figure 2.1 Procurement strategies available. *DBFO and PFI/PPP projects are invariably on a design and build basis

2.2.4 Employer-financed projects

The normal arrangement for construction work is based on finance procured by the project initiator, the employer, client or owner. However, the task of designing and constructing the project can be organised in several different ways. The most common of these are *traditional general contracting* and *design and build.*

2.2.5 Traditional general contracting (TGC)

Here, design is provided by independent or in-house designers (architect, engineer) in direct contract with the client, while a separate contract for the construction of the project is placed with a contractor, who then sublets elements of the work. The typical reimbursement method is a tendered *lump sum*, based on *unit rates* in contractual bills of quantities. The approach is still the most familiar one in the UK and accounts for around 40% (by value) of all projects.

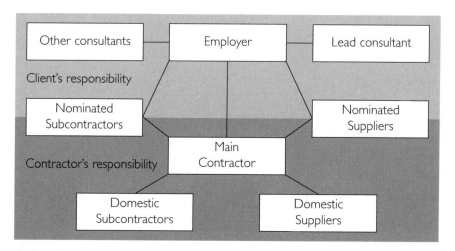

Figure 2.2 Traditional general contracting, showing contractual links

2.2.6 Design and build (D&B)

Design and build (D&B), also known as design and construct (D&C), is a procurement system in which a single organisation undertakes the responsibilities and risks for both the design and construction phases. There may be various levels of employer involvement in the design, but the client must generally give an earlier commitment to the project, together with a willingness to relinquish flexibility and detailed design control. In return, the system can offer significant advantages for the client, including:

- minimum up-front expenditure/in-house resources
- economic design for construction – shorter overall time
- greater certainty of price – reduced risk of extras
- major risks passed to contractor – single-point responsibility.

Despite the disadvantages of higher bidding costs and risks, design and build offers the contractor:

- the potential for almost total control of all aspects, from design to commissioning
- scope for value engineering (see Chapter 5)
- an enhanced opportunity to manage risk in return for reward.

There are many different ways in which design and build is done, as listed below.

True or turnkey design and build
The client engages a building contractor at the outset of the project. The contractor is responsible both for the design and for the construction (see Figure 2.3). This system accounts for about 20% of all design and build projects in the United Kingdom.

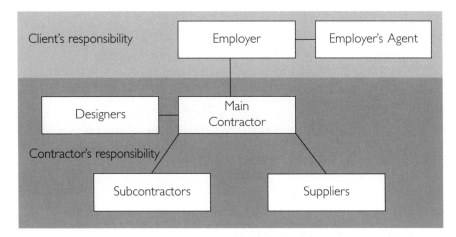

Figure 2.3 True or turnkey design and build, showing contractual links

Novated or 'consultant switch' design and build
The client employs the design team for the early stages of the project (typically up to the planning permission stage) to prepare the outline design and the Employer's Requirement document. The design team is novated to the successful contractor, for whom they then prepare a detailed design. The effect is that much of the employer's traditional design control is retained in the early stages, but ultimate responsibility is passed to the contractor (see Figure 2.4). This is actually the most common form, accounting for about 50% of design and build in the UK. *Novated design and build* has been criticised[1] for restricting the commercial position of contractors, and for creating conflicts of interest for designers, though both groups appear to tolerate the system because of its appeal to clients.

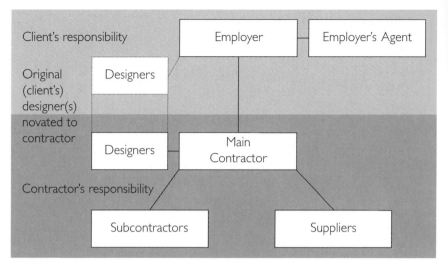

Figure 2.4 Novated (consultant switch) design and build, showing contractual links

Design and develop
The contractor takes on a design produced by the employer's design team and then develops it into a working design, using either in-house or sub-consultant designers. This variant apparently accounts for about 25% of the market,[2] though the precise difference between *developing* a design and *completing* one is not always clear. Indeed, this source of ambiguity can represent a serious contractual risk for the contractor, and has resulted in several notable disputes in recent times.

2.2.7 Management contracting (MC)

The concept of separately procuring a project's management input is based on:

- the changing nature of the construction process, where most of the construction (and indeed some of the design) is executed by specialists under subcontract (Figure 2.5)
- the need for, or existence of, a closer-than-normal relationship between client and contractor, intended to bring improvement in performance, particularly that of time.

The MC contract is normally let on a *cost-reimbursable* (cost plus) basis, with a fee bid for managing the project together with an agreement for reimbursement of expenses incurred. There may be an incentive incorporated to ensure maximum

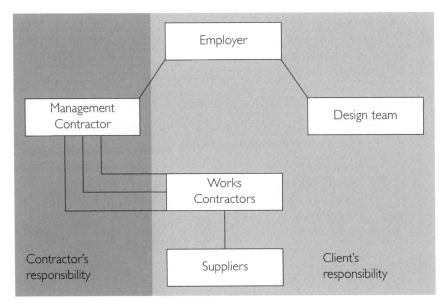

Figure 2.5 Management contracting, showing contractual links

efficiency from the management contractor. MC should be considered particularly where:

- time is of major importance
- the works are complex
- there is a need to start the work before design is complete.

In practice, over the last 10 years MC has been generally replaced by another management-based form of procurement, *construction management*.

2.2.8 Construction management (CM)

Many of the clients for whom MC was originally the preferred procurement method have moved towards construction management (CM). CM was commonly used in Britain in the nineteenth century, is popular in the USA, and is a familiar arrangement in many other countries, for example France. The most significant feature is the direct contractual links between the client and the works contractors (Figure 2.6).

As for MC, the CM contract itself is normally on the basis of a fee bid, an agreement for reimbursement of expenses incurred, and an incentive incorporated to ensure performance. It is used particularly on projects with short lead-times.

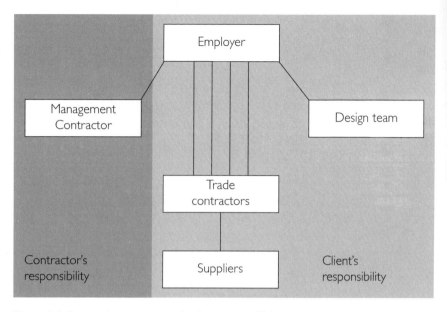

Figure 2.6 Construction management, showing contractual links

2.2.9 Contractor finance: DBFO and PFI/PPP

The terms used for such projects include *design–build–finance–operate* (DBFO), *Private Finance Initiative* (PFI), and *public–private partnership* (PPP). Project structures vary, but typically involve a purpose-made company, called a *special-purpose vehicle* (SPV), which is made up of funders, contractors and operators. This company will

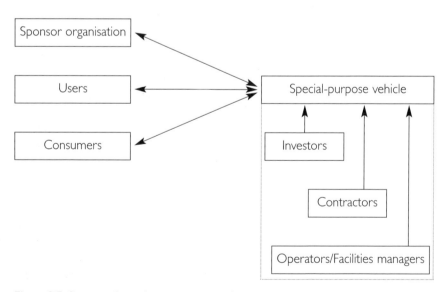

Figure 2.7 Contractor-financed contracts: an example structure

have an upstream contract with the owner, client or sponsor and downstream contracts with constructors, suppliers and service providers (Figure 2.7). In some cases there are additional long-term contracts with the users of the service offered by the project.

2.3   Risk, value and procurement: a summary

Construction projects are complex, and are performed over long time periods under conditions of some uncertainty. There are considerable risks that need to be addressed. The chosen system of project procurement has a major impact on the *management and allocation* of those risks. This is shown in Figure 2.8.

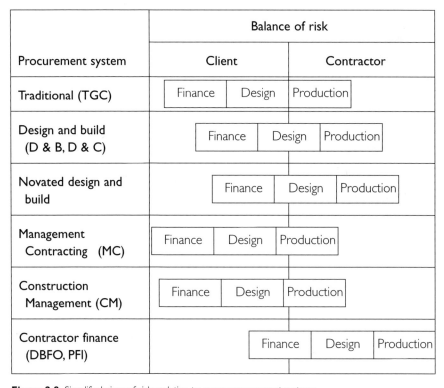

Procurement system	Balance of risk		
	Client	Contractor	
Traditional (TGC)	Finance	Design	Production
Design and build (D & B, D & C)	Finance	Design	Production
Novated design and build	Finance	Design	Production
Management Contracting (MC)	Finance	Design	Production
Construction Management (CM)	Finance	Design	Production
Contractor finance (DBFO, PFI)	Finance	Design	Production

Figure 2.8 Simplified view of risks relative to some procurement systems

The project's procurement strategy is also a crucial factor in obtaining the *required value*. In Chapter 1 it was argued that the aim of value management is to reconcile the conflicts of the project's first-line objectives – time, cost, quality, safety, and environment. Table 2.1 shows how the various procurement strategies support differing client criteria.

Table 2.1 Simplified correlation of client criteria and different procurement systems

Procurement criterion	TGC	D&B	ND&B	MC	CM
Quickest time	*	***	**	**	**
Time certainty	*	***	***	**	**
Cost certainty	*	***	**	*	*
Prestige quality	***	*	**	**	**
Complexity	**	*	*	***	***
Responsibility	*	***	***	*	*
Risk transfer	**	***	***	*	*
Buildability	*	***	*	**	**

***Ideal for criterion; **not particularly ideal for criterion; *not ideal for criterion

2.4   Risk, value and contracts

Imagine a world where all business was conducted by one enormous organisation that designed and delivered products that it had commissioned for its own needs. In that world, contracts would probably not be necessary. However, the real world – particularly the world of construction – could not be more different. We have already considered the *fragmentation* of the construction industry, not just in the way we design buildings and produce them, but in the fact that it is a *project-based* industry and therefore fragmented in time. Such an industry places a heavy reliance on contracts. Contracts are what traditionally have bound together the temporary project coalitions that produce buildings.

2.4.1   What are contracts for?

It is not difficult to see the relationship between contracts and risk. Risk management is a common theme in construction management literature because of construction's complex and uncertain environment. This is a topic that will be covered later in the chapter. But what *value* is added by contracts? To clarify this we must first examine what contracts are supposed to be for.

Contracts between businesses have evolved to take on various roles:

- recording the deal that has been agreed and the rights and obligations of the parties
- providing sanctions for non-compliance, or incentives to comply
- offering sets of procedures that the parties should follow[3]
- catering for uncertainty by deciding in advance how parties will bear the risks of unforeseen events.

Of these four roles, the first three are probably more obvious. The value of having a contract is principally the certainty of knowing what has been promised, how it should be delivered, and what should happen if the promise is not delivered. The more certain a contract can be in these matters, the more valuable a device it is. But what of the role of the contract as an allocator of risk?

2.4.2 Contracts as allocators of risks

Contracts for instant transactions can be relatively simple. They merely define what the performance of the parties should be, and back it up with some sanction. But transactions that take place over a long period of time, such as building projects, require something more from their contracts. One of the principal functions of such a contract is to commit the parties over a period of time.

Classical contract theory works on the basis that a transaction is determined at formation; circumstances do not develop beyond what was envisaged by the terms of the original deal, and so the transaction can be fully planned. In fact, this is often limited by lack of information that would be either costly, or in some cases impossible, to acquire at the time. This is particularly true of the long, complex, exchange transactions that typify modern market activity. This very fact makes *risk* and *uncertainty* indispensable elements of the modern view of contracts.

There can be many uncertainties in long-term contracts: input prices, political market and physical conditions, the attractiveness of the original deal, and even the people involved can all change. (The sources of construction risk are covered in Chapter 1.)

What can also change is the relative power of the parties to cope with or to capitalise on these changed conditions. This is what the risk-allocating function of a contract addresses: it *freezes* the relative power of the parties at a point in time.

Example

A builder is in a hurry to obtain bricks, and places a supply order with a brick merchant. Later, he finds he can get bricks cheaper from another merchant. Thus the builder has just improved his market power by discovering the cheaper price. But the point of the supply contract was to freeze market power at the time of its creation. The builder *could* change suppliers, but there would be cancellation charges to pay. The situation could have been the other way round. The merchant might find a better buyer but he has committed to the builder.

This *freezing of power* is what enables the builder and the merchant to strike a deal in the first place; and the longer or more complex the transaction, the more useful this is. It is this aspect of contracts that causes many authors to refer to them as *risk-allocating devices*.

As we shall see in Chapter 6, there are some fundamental principles of efficient risk allocation. To borrow an expression from the insurance industry, a particular risk should be allocated to the *superior risk-bearer*. In other words, a risk should be allocated to the party that:

- is in the best position to bear that risk; or
- wants to bear it (for some commercial or technical reason); or
- has the greatest incentive to manage and mitigate the risk.

Although many people see risk as a threat, for others it is an opportunity. One of the basic mutual gains that comes from doing business is the opportunity to shift risk from one party to another, depending on their risk attitude. Think of insurance, or betting, for example.

This means that parties will normally choose the contractual risk allocation that best suits their purposes, but, if they wish, they are free to negotiate some alternative risk allocation. The contracting process *should* involve an open and rational negotiation of the ideal amount of risk to accept, or to transfer, given each party's position, their risk attitude, and the relative costs of reducing risk.

2.4.3 Risk allocation: industry-standard contracts

Industry-standard contracts, such as the well-known JCT forms, contain various clauses that relate to risk allocation: for example Clauses 24, 25 and 26 of the Standard Form of Building Contract. Most building projects rely on timely completion, but which party should carry the risk of a delay to the completion date? Generally speaking, both parties have a risk of incurring a loss as a result of such a delay. Their risks can be summarised as:

- Type 1: the loss of use of the building (client's loss)
- Type 2: the loss resulting from continued presence on site (contractor's loss).

Clause 24 of the JCT Standard Form (Damages for non-completion) deals with the client's loss through the device of liquidated and ascertained damages that are payable to the client for *culpable* delay (at a rate stated in the appendix to the contract). The contractor, being culpable, pays for the client's loss, and stands his own. As a result the contractor carries the risk of delays that are of his own making.

Where the delays are *excusable*,[4] however, the contractor will be entitled to an extension of time under Clause 25 of the contract. Delays such as these are sometimes referred to as *neutral events*, since neither party is responsible. Such an extension of time would protect the contractor from the imposition of damages for the extended period (Type 1 losses), but would not reimburse him for any of his own losses (Type 2 losses). As a result the risk of such delays is shared between the parties. Each stands its own losses.

A third possibility, however, is that the delay was neither culpable nor merely excusable, but was caused by some act or omission of the employer (or his agents). We might refer to this type of delay as *compensatable*. In this case the contractor is entitled not only to an extension of time under Clause 25 of the contract, but also to a claim for loss and expense under Clause 26. As a result the risk of such compensable delays falls entirely on the employer, who is, by definition, responsible for the cause of the delay.

Figure 2.9 is a simplified analysis of the above, and represents the way that most standard contracts treat a delay and its consequences.

Source / Impact	Employer	External	Contractor
Employer's loss	E. stands his own loss by giving Extensions of Time	⟶	C. pays E's losses in Liquidated Damages
Contractor's loss	E. reimburses C's loss	⟵	C. stands his own loss

Figure 2.9 Simplified analysis of delay and its consequences

2.4.4 Amending standard contracts for risk allocation

If the risk allocation provided by a standard form of contract does not suit the particular purposes of the parties, they can agree alterations that are more appropriate.

There is strong evidence that standard forms do get amended. A survey commissioned by the JCT[5] revealed that 70% of the employers surveyed had used standard forms, whereas the remainder had drafted their own. However, of those using JCT80 (then the most popular standard form), 72% had made amendments of some kind. The amendments were mostly minor, but were normally because a particular employer found that that standard allocation of risk was unsatisfactory.

Example
The most common example was adjustment of the balance of risk for delay by deleting Relevant Event number 10 in Clause 25.[6] The original clause reads as follows:

1. The Contractor's inability for reasons beyond his control and which he could not reasonably have foreseen at the Base Date to secure such labour as is essential to the proper carrying out of the Works; or
2. The Contractor's inability for reasons beyond his control and which he could not reasonably have foreseen at the Base Date to secure such goods or materials as is essential to the proper carrying out of the Works.

Among the 72% of JCT-based projects for which the standard form was amended, the deletion of this clause was the most frequent amendment. Evidently employers were not content to share the risks of the unavailability of labour and materials, believing this to be a risk that should be allocated squarely to the contractor.

As far as subcontracts are concerned, around a third of the forms in use were standard ones such as DOM/1 and DOM/2, another third were amended standard forms, and the remainder were subcontracts devised by main contractors themselves.[7]

2.4.5 Contractual risk allocation, conflict, and statutory intervention

On the face of it, when parties want to alter the balance of risks between them they agree amendments to standard forms by mutual consent. But the reality is somewhat different.

Example

In a project undertaken by Rayack Construction for The Lampeter Meat Company Ltd,[8] the contractor entered a standard JCT building contract. However, the payment terms were not so standard. Instead of the usual 14 days, the final date for payment had been entered by the employer as 90 days; the defects liability period (normally 6 or 12 months) was 5 years; and the retention percentage (normally less than 5%) was a staggering 50%. Was this risk allocation by mutual consent or a case of the employer taking advantage of its stronger bargaining position or the naiveté of the contractor?

Contractual issues such as delay damages, performance bonds, retention and set-off are key areas for the exertion of this type of *contracting power*. Nowhere has this been more the case than with subcontracts. General concerns about risks in subcontracts have been expressed for many years, though at first this was relatively unspecific: the Banwell Report[9] noted rather tamely that these contracts 'are not free from contention…'

Authors have noted that the contract gives the contractor 'the ability to … control the quantum and timing of payments' to subcontractors,[10] whose contractual position is weaker,[11] and that this is inconsistent with the importance of some subcontractors' input.

Allegations about unfair risk allocation in subcontracts were common in the trade press of the 1980s and 1990s, and various investigators found that subcontractors perceived contracts to be their most critical risk. This received attention in Sir Michael Latham's interim report, which criticised the abuse of power in the procurement process.[12] In his final report Latham stressed the need for statutory enforcement of fair payment terms, and this resulted in the provisions of Part II of the Housing Grants, Construction and Regeneration Act 1996. Among other things, the Act provides:

- a statutory right to payment by instalments for contracts exceeding 45 days
- that all relevant contracts must contain an 'adequate mechanism' for payment
- that no set-off may be made unless prior notice is given, stating the amount and grounds
- that all contracts should provide a right to suspend work for non-payment
- that pay-when-paid provisions are inoperative, except in the case of the insolvency of the employer
- that any party to a construction contract has the right to adjudication.

2.4.6 Contracts: do we need them?

In the early 1960s Stuart Macaulay, an American academic, interviewed businessmen and lawyers from 48 corporations and six legal firms about the way in which they negotiated and enforced their contracts. To his surprise he discovered that many, if not most, of their exchanges had no contract planning, and that 'contract and contract law are often thought unnecessary'.[13] There was also the feeling that detailed contracts got in the way and led to lack of trust and loss of flexibility. Macaulay's study was the birth of an approach to business deals that plays down the role of the formal contract and instead emphasises the *relationship* between the parties. This approach, referred to as *relational contracting,* has recently become of interest with the trend to partnering. One argument can be summarised as follows: if we trust each other, understand one another, and work towards a win–win result, a contract will only get in the way. Partnering, its value and its risks, and whether contracts are required with partnering, will be covered in Chapter 9.

2.5   Conclusions

An understanding of the different ways in which the services of the project participants are procured is crucial for an appreciation of the issues of value and risk. Proper procurement decisions are fundamental to risk and value management.

The following are the key points covered in this chapter:

- In construction projects there are considerable risks that need to be addressed, and the chosen system of project procurement has a major impact on the management and allocation of those risks.
- The project's procurement strategy is also a crucial factor in obtaining the required value.
- Contracts have various roles, including recording the deal, the rights and obligations of the parties, sanctions for non-compliance, and sets of management procedures.
- One of the main functions of a construction contract is to allocate risks between the parties.
- If the risk allocation provided by a standard form of contract does not suit the particular purposes of the parties, then they can agree alterations that are more appropriate.
- Amendments are common, and are sometimes accused of involving unfair risk allocation.
- This has the brought statutory limits in the form of Part II of the Housing Grants, Construction and Regeneration Act 1996.

Notes and references

1 J. Bennett and E. Pothecary *Designing and Building a World Class Industry* (Reading Construction Forum, Centre for Strategic Studies in Construction, 1996).

2 Ibid.

3 For example, in matters such as application for payment or the resolution of disputes.

4 The clause defines precisely what is *excusable* through the so-called Relevant Events in Clause 25.4.1-18 of the Joint Contracts Tribunal (JCT) Form of Building Contract, 1998.

5 Joint Contracts Tribunal *The Use of Standard Forms of Building Contract* (RIBA Publications, London, 1989).

6 JCT98, Clauses 25.4.10.1 and 25.4.10.2.

7 D.J. Greenwood *Contractual Arrangements and Conditions of Contract for the Engagement of Specialist Engineering Contractors for Construction Projects* (CASEC Publications, London, 1993).

8 See *Rayack Construction* v *The Lampeter Meat Company* (1979) 12 BLR 30.

9 Ministry of Public Building and Works *The Placing and Management of Contracts for Building and Civil Engineering Work* (Sir Harold Banwell, Chairman) (HMSO, London, 1964).

10 M. Ball *Rebuilding Construction: Economic Change in the British Construction Industry*, p 151 (Routledge, London, 1988).

11 P.M. Hillebrandt *Analysis of the British Construction Industry*, p 62 (Macmillan, Basingstoke, 1984).

12 Sir M. Latham *Trust and Money: The Interim Report of the Joint Government/Industry Review of Procurement and Contractual Arrangements in the UK Construction Industry*, pp 25–30 (HMSO, London, 1993).

13 S. Macaulay 'Non-contractual relations in business: a preliminary study' *American Sociological Review*, Vol. 28 (1963), pp 55–67.

3 Risk and value in project design

3.1 Introduction

In Chapter 2 we considered how the stage is set for the construction project, in terms of the participants in the process and the way their services are procured. Bringing any project to a successful conclusion is the concern of the many parties that are engaged in its design and construction. A particular difficulty lies in consolidating the project decision-making process when there are so many project participants. In this chapter we examine the process of design; how it adds value; and how that value can be put at risk by a number of factors. In particular, we look at:

- how design teams are organised
- how design teams behave
- how design teams are managed
- the commercial context in which design services are delivered.

3.2 The journey from idea to building

Construction projects are a long time in their realisation. Many different hands will touch a project on its journey from conception of the brief to completion of construction. Inevitably, such a long and complex process will be inherently risky – if only because of changes in the commercial and economic context that have occurred since the original decision to embark on the project. In such a lengthy process there is also a risk of losing sight of the original objectives set for the development.

As the form of the building begins to take shape, questions will inevitably begin to emerge. Indeed the act of designing calls into question both *what* is being designed and *why* it is being designed in this way. What is it that the building is to do, and does it do it? Are some of the things asked for in the brief really necessary? Did we question the original objectives of the brief rigorously enough? Just exactly why is this building being built? What these questions are really asking is:

Does the building add value, and does it add as much value as it realistically can?

Value engineering is a management tool that sets out to examine these questions systematically. Its use is not restricted to construction projects, but because of the length, complexity and uniqueness of each building project it is a powerful and valuable tool for the building industry. It is, however, viewed by many as a redundant process: 'Why do this? Surely if the building is well and properly designed all these matters should have been considered and addressed by the designers?' There is much truth in this, and if the building design process was perfect, then it would be absolutely true.

However, the way buildings are designed in the UK construction industry is far from perfect. The process has many unusual characteristics, many of which can allow or even make inevitable a loss of value in the finished product.[1] It is worth examining the design process, both to understand how value management can contribute and to explain why value management might have something to offer. Value management is not a universal panacea. It is not a substitute for bad design, nor for badly managed design, nor for that matter for bad designers. Formal value management is probably not necessary or cost-effective for smaller projects.

3.3   Participants and the process of building design

The construction industry is often considered to be unique. One particularly unusual feature of construction in the UK is the split between the design and construction processes in traditional building procurement. The management and organisation of the participants in the pre-construction design stage can act as a negative force militating against arriving at the optimum solution, rather than a positive force for adding value. There are several issues that need to be examined in order to understand this:

- the way in which design teams are organised
- the way in which design teams behave
- the control systems used in design management
- the commercial context in which design services are delivered.

3.3.1   Design team organisation and its impact on value

Discontinuity: the forming, storming, norming and performing problem
One of the unusual characteristics of building design is the discontinuous nature of the process. A building is usually designed by a team brought together specifically to work on that particular project – what in management theory would be described as a 'temporary project coalition'. On completion of the project the team is disbanded, possibly never to work together again. This creates both organisational and behavioural team-working problems. The typical life of a project team goes through four stages: *forming, storming, norming* and *performing*, followed (in project-based teams) by the *mourning* – in other words, the disbanding of the team. Figure 3.1 illustrates this.

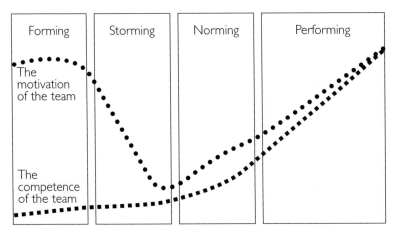

Figure 3.1 Influencing a project's cost outcome: ability and cost

During the *forming* stage the team members are learning about each other's strengths, weaknesses and personality traits, and developing an organisational framework within which to operate, setting up practical management systems for information exchange, meetings, and project budgetary control. At the same time the project is going through the crucial early inception stages. The design team is seeking to understand the nature of the project and the nature of the user organisation, to develop the brief and carry out feasibility and viability studies. Decisions made at this stage can have important long-term implications for the project, and are difficult to undo. If the project sets off on the wrong foot, value can erode away and matters can be difficult to correct further down the line.

The ability to influence the cost of a project decreases as the project progresses. At the same time the cost of implementing changes rises, both in direct terms and indirectly in terms of the cost of redesign work and programme disruption.[2] Figure 3.2 illustrates this point.

The *storming* stage of teambuilding is characterised by creativity and conflict. By the *norming* stage the team has reached an accommodation and has developed ways of working together, both at an interpersonal level and in terms of management procedures and control systems. This improves efficiency, which steadily increases into the *performing* phase, the period of maximum working efficiency. At the *mourning* stage the team is disbanded. Depending on how this is done the shared accumulated learning, the organisational working systems and the team *esprit de corps* can all be lost. Capturing this knowledge base and allowing opportunities for continuity of team working can yield great future benefits. In the traditional process this rarely happens, and one of the key aims of strategic partnering (as will be discussed in Chapter 9) is to hold on to this accumulated experience.

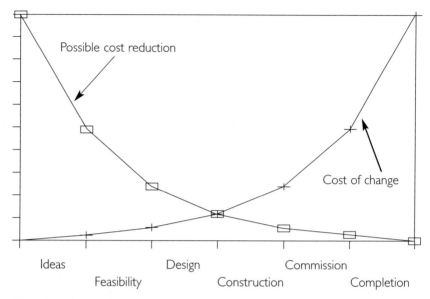

Figure 3.2 The cost of change increases as the project progresses

Value management is typically carried out at the beginning of the *performing* stage. By this stage more information about the project should be available, and the design team should have acquired a fuller understanding of the client's aspirations. Value management can contribute by reviewing the decisions made during the shaky *forming* stage and testing the quality or validity of earlier decisions. Smoother ways of working together and closer interpersonal relationships should have developed, and the team will have a more open attitude to exploring and brainstorming ideas for improvement.

The fragmentation problem
Fragmentation occurs both between the design and building functions and within each of these two areas. Fragmentation in the building industry is caused by the splits between contractor, subcontractor and supplier, with further splits caused by labour-only subcontracting. (Fragmentation in the on-site construction process is dealt with more fully later in this chapter.) At the design stage it is exceptional for construction consultancy companies to be multidiscipline. Engineers, architects and surveyors traditionally operate in separate organisations. This provides important commercial benefits: a high degree of flexibility, and the opportunity for companies to acquire and provide specialist expertise. However, it also creates organisational difficulties – all the communication, management and control issues that occur in any temporary grouping of people.

The design and build split is further complicated by the fact that it is a market rather than a hierarchical split. The gap must be bridged by a commercial transaction – the act of tendering. The lack of easy access to either side of this commercial barrier makes it difficult or even impossible to share information and knowledge between builder and designer. This is why it is notoriously difficult in traditional procurement to increase value by improving buildability.

The 'over the wall design' problem
The traditional sequential design process involves what in manufacturing has been described as *over the wall design*. The client prepares the brief and throws it over the wall to the architect and engineer, who prepare layouts and specifications and throw these over the wall to the quantity surveyor for costing. Typically the whole thing is thrown back over the wall to the client at that point, when it becomes apparent that the building as briefed cannot be built for the budget allocated. This iterative process is clearly profligate of time – and therefore of cost – and is also unlikely to throw up the best-value design solution.

One initiative pursued to overcome this problem is *concurrent engineering*. This is based on the concept of taking the design forward across a number of fronts at the same time, rather than using the traditional sequential approach. The car manufacturing industry has had some success in reducing design time and improving quality by breaking down the traditional barrier between design and production and encouraging a concurrent, rather than sequential, approach to design.

A further common 'over the wall' problem occurs when part of the design is to be carried out by a specialist subcontractor with design responsibility – for example where the JCT Contractor Designed Portion is used.[3] Typically a performance specification is produced, and this is used as the basis of the contractor's detailed design. The detailed design is prepared out of sequence, which may lead to compromises in the design or the need for other elements of the building to be adapted to accommodate the element as designed. A common example is the mechanical and electrical installation in a building, where the requirement to provide the air changes to meet the performance specification may lead to duct sizes that affect ceiling levels or the zone for the structural elements. Such out-of-sequence working may make commercial sense in the short term, but in the long term it is unlikely to create the best-value solution.

All these issues, which anyone familiar with traditional contracting will recognise, militate against achievement of the best-value solution. The unique nature of the pre-construction design stage, and the transitory relationship and organisation of the participants, can create an environment in which value trickles away. Good design management can remove or at least minimise these problems. Value management can contribute to breaking down the barriers and bringing together all the parties early in the process. A formal value management review process is an important component of the wider design management system.[4]

3.3.2 Design team behaviour and its impact on value

Revealing the bigger picture
A common problem in design management is project managers' failure to reveal the bigger picture to the complete design team. Typically the senior designers and managers will attend project design team meetings where strategic issues are reviewed. They then feed back information and define tasks for their internal design team. The obvious problem with this is the quality of the feedback. Often individual tasks are well defined; the more common problem is a failure to reveal the bigger picture. This can have a significant impact on value: although the technical issues may be resolved at a detail level, the bigger question of how these details add value to the total project is not addressed. This process also fails to capture the experience, knowledge and skill of individual team members, which could be of enormous value. John Worthington sounds a warning on the tendency in management thinking to lose sight of the bigger picture:

> The power of management thinking is the ability to tame 'wicked problems' by subdividing complex problems into their more manageable component parts. The failing is that whilst the parts are manageable, the coherence of the whole is lost. The strength of design education is to instil a way of thinking that leaves options open – finding form out of chaos and presenting holistic solutions.[5]

Value management can allow a review of the bigger project objectives in a way that allows a cross-section of the entire design team to be involved.

Professional responsible autonomy
A related issue is the need to allow a degree of professional responsible autonomy within the design team. There is a constant tension in construction design management between the need to manage the design process towards the project objectives (particularly budget and time constraints) and the need to allow individual designers the time and freedom to explore alternative solutions. Management that simply focuses on outputs will miss the opportunity to capture individual professional skill, and will also lead to a demotivated design team. Failure to strike a balance between individual professional motivation and the needs of the project will inevitably lead to a loss of value.

Client relations
The relationship between the design team and the client is important, and can impact on value. It is important to establish that the designer is a professional adviser, not just a supplier of design services. If the design team is to add value then it is important that their advice is proactive. As with all professional advisers (think of doctors and accountants) they may well be giving advice that the client doesn't wish to hear. That of course doesn't mean it is bad advice. Often design

teams accept the brief and provide a built form solution that simply translates operational requirements into a building. More value is added where the design team use their own knowledge base to question the client's needs and provide solutions that exceed rather than simply meet the client's aspirations.

Standard contracts: contract administration and other procedures
A standard contract in the construction industry is both a legal agreement that sets down the intentions of the parties and an operation manual that defines how to deal with matters that arise during the contract. In the traditional arrangement a member of the design team, usually the architect, is responsible for administering the contract between the employer and contractor. This arrangement has its advantages, but it also has disadvantages that can inhibit the adding of value. The contract defines strict roles for designer and builder that discourage integrated working. Standard contracts encourage automatic responses to problems, based on habit and defined rules. These procedures tend to be framed in a way that avoids bad outcomes rather than maximising good solutions (those that add most value).

Example

During the refurbishment of a listed building the existing roof construction was uncovered. It was discovered that the layout of the existing timber structure would allow the air conditioning plant to be positioned within the roof space rather than within one of the top floor rooms. It had not been possible to establish this before the building works started. Making this change would increase the usable floor space of the building and hence increase the value of the building, both in financial terms and to the operator. The standard contract under which the works were being carried out did not easily allow such a change to be implemented. It was theoretically possible for the client to make the change, but the terms of the contract relating to cost and time were such that in practice it was easier to leave the design as it was. There are good reasons why standard contracts are framed in this way; however, as a result they make it difficult to take advantage of windfall opportunities to increase value.

3.3.3 Design management and its impact on value

Building design would be considerably easier were there only one customer for the design, but this is seldom the case. Usually there are multiple customers for the design, often with conflicting needs. For example, the M&E engineer will require adequate space for a plant room taking into account maintenance and installation access. This will have to be balanced by the developer's desire to maximise the usable or net lettable area that contributes to their income stream. Value management can play an important part in providing a forum in which balanced decisions can be made.

This constant trade-off between competing design criteria happens throughout the design. It is further complicated by the fact that this 'trading' often takes place with inadequate information and under intense programme pressure.

Early design is particularly susceptible to risk, in that it is hard to evaluate and control against time, cost and quality standards. These risks – of failing to achieve what was intended – are a function of the constant project uncertainty that surrounds most construction projects. This uncertainty arises from such things as:

- pressure to reduce time and cost
- technological changes: different ways of making the building
- shifting client requirements.

For example, during the design of a new store a retailer-client may face sudden new market demands created by changes in consumer behaviour. They may find that there is greater demand for fresh products and less demand for frozen foods. This will have an impact on the building design in terms of the store layout, the M&E plant, the budget and so on.

Risk management provides a forum in which to identify and assess these changing needs at an early stage, and value management provides the solutions to the problem of how to reconcile them.

Many of these organisational, management and behavioural issues are of course exactly what new forms of procurement, and partnering in particular, seek to address. The way value management differs in this new context is dealt with in Chapter 9.

3.3.4 Commercial issues and design services: their impact on value

Professional services in the UK construction industry are provided largely by private independent companies. Since the abolition of mandatory fee scales, fees are often negotiated in competition. As with tendering for the actual construction works, so clients increasingly seek competitive bids for the design and project/cost management services. In such a tight commercial environment consultants are increasingly careful to define what is and what is not included as part of the service to be provided. Profitability for the consultant is preserved by ensuring that the actual service delivered stays within these boundaries.[6]

Clients view value management with some suspicion. First, they would tend to take the view that an important and inherent part of the design process is the constant pursuit of the best-value solution. They may well claim, with some justification, that it is not something that should be seen as an optional extra, and certainly not something they should have to pay additional fees for. Second, and again not unreasonably, they require clear evidence that they will receive a return on their investment in commissioning a formal value engineering study. Anecdotal evidence suggests that value management can lead to cost savings of 10%, but it is difficult to produce hard figures to support this.

It has been said that the three things required for good design are time, time and time. However, in general there is considerable pressure to reduce the design time on construction projects. For the client, the time between site acquisition and occupation and use of the building is a period when costs are incurred with no income. For the design team, it is generally true that the quicker the design is complete the more profitable the commission will be. These commercial pressures make it unlikely that anything more than the minimum time will be given to exploring, testing or validating design options in terms of added value.

There are concerns beyond the commercial that will motivate designers to pursue value. There are legal obligations – the duty of a professional practitioner to exercise skill and care; in theory at least this would include pursuing (although not necessarily achieving) a solution that is not profligate of the client's money. More important, however, is the professional's moral obligation to provide value for money. In structural engineering, for example, there is a strong tradition of pursuing economic elegance in design. These are very important motivations.

However, value management has an important contribution to make in allowing a *formal* review of the design. It provides one of the few opportunities to audit and question the design during the pre-construction rush to site.

3.4 The design process: the traditional model

3.4.1 Issues that put value at risk

The traditional process of procuring buildings follows a linear progression. The design team, usually led by the architect, is appointed at the outset to take the client's brief, to carry out a feasibility study and to advise on procurement of the building. This is followed by preparing and submitting applications for the various statutory approvals required; preparing working drawings, specifications and a bill of quantities; and obtaining tenders. It is only at this point that the contractor and subcontractors – those who will actually construct the building – appear on the scene.

By this time the design is set, not only in concept but also in some considerable detail, and opportunities to take on board the contractor's input to the design are severely restricted. The proposed development has also by this time usually received third party approvals (planning permission and possibly building regulations approval), and any changes beyond the most minor will require resubmission and further approval. As there is no certainty that such approval will be forthcoming, it is difficult to revise the design significantly after this stage, and the process of reapplying for approvals is both risky and time consuming.

Attempts to incorporate the input of the downstream suppliers or subcontractors are possible but only either by the unwieldy mechanism of nominating specialist subcontractors or by informal pre-tender consultation together with a specification written in such a way as to effectively remove other choices. In practice, both are fraught with difficulty. The administrative and contractual complications involved in nomination have made it so unpopular as to be virtually non-existent in current practice.

In this traditional process the design team remains alongside the client and administers the terms of the building contract from start on site through to completion. The process is well established and well understood – certainly by construction people if not by their clients. The RIBA Plan of Work[7] is an industry standard, which describes and defines this process clearly and thoroughly. For many years this traditional form of procurement was the only game in town – nearly all buildings were procured in this way.

Notwithstanding this popularity, the process was and is fraught with difficulties, and is in many ways inherently inefficient, not least in the way it splits the design and construction functions. It has many advantages, but it also has many characteristics that tend both to increase risk and to reduce opportunities for adding value. These occur particularly in the early stages during the briefing and early design activities.

3.4.2 Starting out: the brief

Good design gives rise to good buildings. Similarly, good briefing creates good design and therefore good buildings. It is impossible to make significant improvements to the construction process without rethinking the design. Similarly, it is impossible to improve the design without addressing the brief.[8]

The basis for the design and the eventual building is the client's brief. It is vital to get this part of the process as right as it can be. Time spent on briefing is rarely time wasted. In general too little time and effort is put into this stage, and this can create acute problems later in the design and construction. There are particular problems that arise from the briefing process that can have a significant impact on risk and value.

The brief can be considered as having two distinct component parts: matters to do with the product, and matters to do with the process. Product issues are essentially concerned with *what the client wants from the building* or *the building the client wants.* Process issues relate to *how the building is to be delivered.*

Defining the process
At the outset it is important to define the level, nature and scope of service to be provided by the design team. It is then necessary to set criteria for time, quality and cost:

- How much will risk be dealt with by design and how much by procurement? How is risk to be dealt with – by managing, avoiding, sharing, or transferring?
- How will cost and value management be dealt with in the process? Change is inevitable, but how is it to be dealt with? What about changes arising from the value management review? What about changes that reduce risk but also reduce value – who will decide on the preferred course of action, and how will this decision be made and agreed? How will the client, user or owner be involved in this consultation process? How will conflicting aspirations be resolved?
- What information management procedures will be put in place? How will control be exerted to ensure that value doesn't slip through? Will there be a single point of client contact or will access be available to the bigger organisation?
- Will the whole construction supply chain be involved and, if so, how and on what basis? There are opportunities both to add value and to reduce risk by involving specialists in the early design stages, but will the procurement system allow this? If not, then should the procurement route be changed?

The way in which the project management systems are set up at the outset can have a fundamental impact on the ability to manage value and risk. It is important to set off on the right foot by developing a management framework that recognises and addresses these issues. Ideally the brief should define not only *what* is to be built but also *how* it is to be built.

Defining the product
This part of the brief is concerned with defining the building. The amount and degree of detail contained within the briefing document can vary enormously. At one end of the scale, the brief can set detailed prescriptive standards for the building – accommodation schedules, room services data, and so on. At the other extreme the brief can be a loose, open document that provides the minimum performance information – the use of the building, the total area required, and so on. This of course should not be confused with abdication of the responsibility to produce a brief at all.

There are advantages and disadvantages to each approach. A prescriptive brief may constrain the design team and lead to missed opportunities to add value by imaginative design. A performance brief may result in a lengthy and unproductive design process that ultimately gives rise to a building design that fails to answer the needs and aspirations of the client.

Defining needs or outputs
The fundamental function of the brief is to define the operational requirements of the building. At its most basic this would usually extend to setting down:

- the areas required to accommodate the various functions
- the uses or functions that will take place in the individual areas
- the relationship of functions and areas to each other (there may also be critical relationships that need to be accommodated in adjoining spaces, or juxtapositions of uses that must be avoided)
- fixtures, finishes, equipment and environmental performance (for example lighting levels, temperature, humidity).

To be able to add value to the project by design it is necessary to understand not just the operational brief as presented but also the logic behind it.

Defining aspects of the product: quality
It is often very difficult to define 'quality' in the construction process. And as quality and value are interlinked, confusion over one can have serious implications for the other. There is a fundamental problem in the construction industry inherent in the tradition of separating *time* and *cost* from *quality*. Any normal measure of quality would include, as key measures of quality, meeting the budget and delivering on time.

The problem is not so much that quality isn't defined, but that there are often too many interpretations of it. For example, *on-site quality* normally means conformity with the contract specifications. *Design quality* is taken by designers to mean the identification of customer needs and their translation into design solutions. But the most common use of the word 'quality' is in the context of the specification. When people talk of a *high-quality building* they usually mean a building built to a high (that is, expensive) specification.

To manage value successfully there must, at the outset, be a clear consensus about the meaning of quality. If quality is defined as the *level of specification,* then this should be clarified by reference to:

- performance criteria
- appearance
- functional attributes.

At the briefing stage it may help both the client and the design team to relate these to examples of completed buildings. The more explicit the definition of quality the more objective the management of value can be.

Defining aspects of the product: cost
Equally, there must be no ambiguity with regard to cost. The following should be considered and clearly defined:

- The capital cost of the development: what this includes and excludes, and who is responsible for what. In particular, what will fall within the construction contract and what will be part of the occupier's fit-out? There is little point in value-managing matters that fall outside the scope of the contract.
- The life cycle costs, the operating costs and their relationship to the capital costs.
- How cost contingency, sensitivity and risk are to be dealt with. What allowance is there in the budget for contingency, and how is this expenditure to be dealt with?

If there is ambiguity over cost it will be very difficult to deal with value issues. If the budget is ill defined, or a 'moving target', this will create a variable in the cost and value equation that is impossible to resolve.

Defining aspects of the product: time
The total duration of the project must be clearly defined – not just the design and construction, but any time necessary for activities that are required for the building to operate, such as commissioning, fitting out and merchandising. Definition of time should also extend to defining critical programme dates. Many buildings have key events that they must meet: for example the start of term for education buildings, or Christmas for retailers. Missing these dates would create a loss that would probably far exceed the additional cost of completing the building more quickly. In considering value this would be a very important factor. A value management exercise in this context may have time as its focus rather than cost or quality.

3.4.3 From feasibility study to tender

This stage of the project consists of two related activities. The first is to test the *feasibility* of the project. This involves examining the physical aspects of the development, and asking questions such as: Will the building fit on the site? Are the ground conditions suitable? Is there an access point for vehicles that will be acceptable to the highway authority? How big does the building need to be to accommodate the process?

The second activity is to examine the *financial viability* of the project. The building promoter – the developer or the occupier – will wish to test the cost of the project in a development appraisal or in a business case, or to apply some form of cost–benefit analysis. In all cases the project will be required to demonstrate a positive and adequate return on investment, or to show that the derived benefits exceed the likely costs. The work carried out by the design team at this stage will fundamentally influence and inform the decisions made. However, it is unlikely that there will be sufficient design work at this stage on which to base a formal value management exercise.

During the development of the scheme design the physical plan and specification will begin to take shape. It is at this stage that a value management workshop is likely to yield the greatest benefit. There is enough design work to form the basis of a review, but not so much that it cannot be reworked. For reasons outlined earlier it is important that the review is carried out before approvals such as planning permission and building regulations approval have been obtained.

As work progresses to the detail design stage the design process will move from considering strategic matters to specifying and detailing the individual elements and components of the building. Unless a decision has been taken to revisit the design and the brief, any value management at this stage would be concerned with:

- ensuring that the original value objectives are not being lost or eroded in the detail design
- checking that adding value in one element of the building is not at the expense of another
- reviewing and value-managing individual elements of the building, including design work by specialist suppliers and subcontractors.

Once the tender documents have been prepared the opportunity to value-manage the project will move to the post-contract stage. This is dealt with in Chapter 4.

3.4.4 Design and build: some examples of value at risk

The cut-off between the pre- and post-contract stages is less clearly defined in *design and build* contracting. Until now we have assumed only one procurement scenario – the *traditional* model, outlined in Chapter 2, section 2.2.5. With this model, the design process follows the RIBA Plan of Work, and the construction work is usually let on one of the JCT Minor Works, Intermediate or Standard forms of contract. Separation of the design and build functions is also common to *management contracting* (see section 2.2.7 of Chapter 2) and *construction management* (Chapter 2, section 2.2.8), although not to the same degree.

Up until the last 30 years this separation of functions was pretty much the only game in town; nearly all projects were carried out on this basis. From the 1970s onwards design and build procurement has increased in popularity, and has grown steadily in market share.[9]

This move has probably been as much customer led as supplier led. It has also been driven more by a desire to simplify the client's relationship with the design and construction team, than to encourage a more intimate design and construction relationship. It has generally not been driven by a desire to increase buildability, but more often has been seen as a means to reduce risk and costs on the part of the client.

The variants of the design and build procurement method are outlined in section 2.2.6 of Chapter 2. At first glance it would seem that design and build, by integrating the design and construction process, should improve the delivery of value. However, it can bring with it many problems that can lead to a loss of value. Some of these are listed below.

The artificial unity problem
Design and build contracting generally involves the same split in organisational terms that exists in traditional contracting. Apart from the very small number of contractors with in-house design departments, the most popular form of design and build involves the novation of the design team. Often the contractor will have no experience of the design team and, worse still, in this 'arranged marriage' will have no say in the suitability or otherwise of the designers. Not only is the marriage arranged, it is also temporary, and will be terminated as soon as the project is completed. There will probably not be any significant opportunities to integrate the design function to achieve greater value.

The timing problem
There is rarely any opportunity for the contractor to influence the design. By the time the contractor is appointed the design is generally well developed. Almost certainly the form and layout of the building will have been decided, and the only influence the contractor will be able to exert will be on the detailed design – the assembly of the building. It is usually possible to influence cost at this stage, but it is very difficult to influence value. So in the most popular form of design and build there are only very limited opportunities to capture the contractor's construction and management skills.

The inflexibility problem
As a form of procurement, design and build does not easily allow the introduction of changes once the contract has been let. Once the contractor's proposal has been accepted, it is very difficult to make changes to the design without incurring

great cost. Indeed it is generally accepted that design and build is not an appropriate procurement method if there is any likelihood of changes to the employer's needs. However, change may come about for reasons beyond the employer's control, and this may lead either to costly contractual variations or to the acceptance of a compromise that ultimately reduces the value of the finished building.

Related to this is the need to ensure that the Employer's Requirements[10] accurately describe the building the employer wants. This is a key document in design and build contracting. It is the contract document that sets down what the employer wishes the contractor to build. There is a dilemma here: the more detailed and explicit the Employer's Requirements, the less opportunity there will be for the contractor to introduce improvements to the design arising from, for example, his technical, purchasing or management ability. The less detailed the Employer's Requirements the more likely it is that the building will not satisfy the employer's detailed brief. This is why design and build generally works well with loose-fit buildings that contain flexible functions. Complex buildings with complicated functional needs are unlikely to deliver the greatest value if procured on a design and build contract.

3.5   Conclusions

The following are the key points covered in this chapter:

Design management problems that may impact on value

- The split between designers and builders
- The temporary nature of project teams
- The problem of 'over the wall design'
- Design and build is not always a solution for delivering *best value*, especially for complex buildings.

Issues to be aware of

- The need for the full team to see the bigger project picture
- The need for a degree of professional autonomy for individual members of the design team
- The need to understand the client's brief fully, and proactively to pursue value in developing the brief
- The need for the chosen procurement route to deliver *value* as well as the more common criteria of *time, cost* and *quality*.

- Involve the production supply chain with the design team early in the project development. This may be effected by a procurement arrangement that allows this. Examples are partnering, two-stage tendering, management contracting, and some forms of design and build. In any event the supply chain should ideally be involved in some way in any pre-contract value management workshop.
- Management procedures should be inclusive, and should involve all levels of participants. Care should be taken to communicate the wider project objectives to the full team. Face-to-face briefings should constantly restate the project objectives and highlight any changes and new information that has emerged.
- The professional knowledge and skill base of individuals should be recognised and acknowledged. It is vital that the project manager encourages individual input and gives this the appropriate weight within the wider picture.
- The client's brief should be seen as a dynamic document to be constantly questioned and revisited as the design and construction develops. Understanding of the client's brief should exist at both a strategic and a detailed level. This may involve spending time within the client's organisation.

Notes and references

1 For those interested in understanding the history and background to this fragmentation see C. Powell *The British Building Industry since 1800: An Economic History* (E & F N Spon, London, 1996).

2 To understand value management it is necessary to appreciate some of the issues involved in managing building costs. This is a large, complex subject that falls outside the scope of this book. There are however many excellent texts on this subject; one of the best and most popular is D.J. Ferry, P.S. Brandon and J.D. Ferry *Cost Planning of Buildings*, 7th edn (Blackwell Science, Oxford, 1999).

3 Published by RIBA Publications.

4 For further reading on managing the design process see C. Gray, W. Hughes and J. Bennett *The Successful Management of Design: A Handbook of Building Design Management* (University of Reading, 1994).

5 J. Worthington 'The changing context of professional practice' in D. Nicol and S. Piling *Changing Architectural Education*, pp 27–40 (E & FN Spon, London, 2000).

6 For further reading on the appointment of professional consultants see R. Byrom *Terms of Engagement and Fees* (RIBA Publications, London, 2001).

7 S. Lupton (ed.) *Architect's Job Book,* 7th edn (RIBA Publications, London, 2000).

8 Briefing is a complex subject, and is dealt with here only in as much as it relates to cost and value management. For further reading on the subject see D. Hyams *Construction Companion to Briefing* (RIBA Publications, London, 2001).

9 For further reading on the subject of design and build contracts see D. Chappell and V. Powell-Smith *The JCT Design and Build Contract*, 2nd edn (Blackwell Science, Oxford, 1999).

10 For a further explanation of these and other building contract legal terms used in this book see D. Chappell, D. Marshall, V. Powell-Smith and S. Cavender *A Building Contract Dictionary*, 3rd edn (Blackwell Science, Oxford, 2001).

4 Risk and value in project construction

4.1 Introduction

Building projects are produced in ways that involve a number of participant teams, who are procured under various systems, and who operate in a number of ways. In Chapter 3 we looked at the implications of this for risk and value in the design process. Now we turn our attention to when the designs are implemented – the construction stage of the project. In this chapter we examine the construction processes that are carried out, predominantly on site by the temporary project teams that have emerged from the procurement processes described earlier. In particular this chapter examines:

- the make-up of site teams and the fragmentation of on-site production
- types of subcontractor
- the value added and risks created by subcontracting.

These issues were illustrated in Table 1.1 in Chapter 1 (repeated here for convenience).

Table 4.1 Risk and value issues classified (repeated from Table 1.1)

Issue	Examples
Political/regulatory	Government. Planning. Contractual. Pressure groups. Health and safety. Environment
Technical and physical	Logistics. Geology. Topology. Weather. Culture and religion
Conceptual	Client organisation. Project scope and definition. Design
Financial	Funding. Tax. Financial stability. Inflation. Exchange rates. Bonds
Organisational	Estimating. Scheduling. Mobilisation. Project structure. Control. Communication.
Operational	Quality. Safety. Innovation. Design change. Labour. Materials. Plant. Subcontractors

4.2 Risk and value, and fragmentation of the on-site process

4.2.1 Fragmentation of the on-site process

Critics of the industry in the mid to late twentieth century tended to concentrate on the design–build fragmentation in the traditional procurement system. More recently, following the trends to integrate these processes (as described in Chapters 2 and 3), the focus for criticism has also included the on-site fragmentation between those responsible for construction.

4.2.2 Subcontracting

By far the biggest element of a project's on-site value is added by specialist and trade contractors, who are typically engaged under subcontract by the main contractor. This situation is not unique to construction, nor to the UK: the significance of subcontracting in other industries (textiles, electronics, food and motor manufacturing) and in other countries (European, American, Asian and Australasian) is well documented. The reasons *why* subcontracting is so extensive will be discussed later, but it is beyond doubt that the relationships between a contractor and its subcontractors are crucial in terms of maintaining or enhancing a project's value and managing the associated risks.

It is difficult to establish the exact extent of subcontracting in the construction industry: complications include the sometimes ambiguous status of labour-only subcontractors, and also the fact that primary subcontractors themselves regularly sublet. Some projects are entirely subcontracted; with others, the main contractor's contribution amounts to a 'skeleton' management team.[1] Overall, and taking into account regional and other differences, a conservative estimate would be that over 60% of a project's value is likely to be sublet by the main contractor.

4.2.3 Types of subcontracting, and the make-up of the on-site team

In construction, subcontracting involves site work, either on its own or with any combination of the three other main value inputs: *design, manufacture* and *supply*. Combinations range from *fix only* to a complete *design–manufacture–supply–fix* package. Sometimes the service extends beyond this, into commissioning and operation. Some people prefer to differentiate between *specialist* subcontractors, who generally bring some design input or advanced technology to the project, *trades* subcontractors, who represent the more traditional skills, and *labour-only* subcontractors, who generally exist because of the main contractor's economic preference for contracting labour rather than employing it directly. These categories reflect two underlying causes of subcontracting (one technical, the other economic) that will be mentioned later.

Another way of classifying subcontractors relates to their contractual status. According to the drafters of the JCT contract, a *domestic subcontractor* is any person to whom the contractor sublets any portion of the works and for whom the contractor remains 'wholly responsible'. In other words neither the employer nor his agents have any influence or responsibility. The subcontract may be an industry standard form (such as DOM/1 or DOM/2), an amended version of the same, or a main contractor's own form.

Example
A domestic curtain walling subcontractor carries out work that proves to be defective. It will need to be removed and replaced, which will cost £500,000 and delay completion of the project by 20 weeks. The consequent disruption will cost the main contractor £250,000 and other subcontractors a total of £100,000. Since the subcontractor is domestic, initial responsibility – for the defect, the delay, his own losses, and those of the other subcontractors – falls upon the main contractor. Of course, the main contractor will seek to recover these losses from the subcontractor.

Nominated subcontracting occurs when the client or his consultant has (again quoting JCT) 'reserved to himself the final selection and approval of the subcontractor to the contractor'. The practice of nomination evolved for four principal reasons:

• an increase in technological complexity
• the desire of clients to enjoy stable business relationships with certain key specialists
• the increasing importance of specialist contractor design and the need for direct client–specialist design links
• modification of the main contractor's control and risk exposure.

Example
If, in the previous example, the subcontractor had been nominated, the situation might have been somewhat different. Although the main contractor would still have been responsible for the defect, his own losses, and those of the other subcontractors, there would be an argument that he was entitled to an extension of time because of 'delay on the part of a Nominated Subcontractor'.[2] This has the effect of transferring some of the risks (those of the delay) onto the employer.

Although there has been a decline in the practice of nomination, some clients have sought to retain influence over the selection of subcontractors through other devices. Some of these, such as *naming* and *specifying*, are incorporated in the JCT's documents. Sometimes clients use looser terminology: in many cases the adoption of *preferred* subcontractors is merely dictated by the client or his consultants, and direct design liability, where required, is ensured by collateral warranty between the client and the subcontractor.

4.2.4 Where subcontracting adds value

The factors that have led to the emergence and growth of subcontracting apply to many industries, but they have been particularly significant in construction. There are three sets of reasons for this: technical, economic and managerial.

Technical value
Building a project requires a broad spectrum of specialist skills and technologies that either cannot be, or are too expensive to be, retained in-house by the contractor. This factor will grow in relevance as buildings become more technologically complex.

Economic advantages
Subcontracting carries great economic advantages for the project. First, there are the cost benefits to the main contractor of switching overheads to other companies, as testified by the current passion for outsourcing in industry generally. Second, there is the issue of supply and demand. The ability to subcontract overcomes problems of capacity in the face of volatile demand. But this is not the end of the story. In such times of economic uncertainty, subcontracting not only affords protection to the main contractor from fluctuating demand, it also allows them to take commercial advantage of the situation. The profitability of many general contractors has been based on their ability to take full advantage of the contracting system (in reality the *sub*contracting system) to wring a profit from the industry's economic cycles and the labour shortages or surpluses that resulted.[3]

Managerial control
Another benefit of subcontracting is that it facilitates managerial control, or at least makes it less expensive. Contractors tendering in competitive markets have had to make do with ever-diminishing managerial overheads. They have compensated for this by increased subcontracting, which permits financial control to replace production control (and production managers to be replaced by commercial managers) as firms no longer manage directly employed workers but bring

together specialist packages. In 1970, the ratio of operatives to administrative staff[a] employed by main contractors was 4 to 1; by 1985 it had reduced to 2 to 1, and by 1996 the ratio was 1.5 to 1.

4.2.5 Where subcontracting creates particular risks

Despite its advantages, subcontracting can create risks for the project as a whole, and far from improving efficiency or profitability, these risks carry the potential for the opposite effect. The potential risks from subcontracting can be summarised as follows.

Control
Subcontracting substitutes firms for directly employed workers, and thus substitutes financial control for production control. This has two effects:

- a reduction in the willingness (and ability) of the main contractor to intervene in issues concerning the subcontractors' management of their own time, cost and quality
- an attitude of 'compartmentalised responsibility' between individual subcontractors.

The ultimate result is that time cost and quality are put at risk.

Coordination and communication
Although it is convenient to break the project down into separable, subcontracted work packages, the interfaces between separate specialist packages are a constant source of risk to the main contractor, since it is not always clear where responsibilities lie between subcontractors. Often these interfaces are neglected, or incorrect assumptions are made.

Health and safety
The responsibility for health and safety issues on-site lies with the principal contractor, but this demands vigilance in making sure that individual subcontractors (many of whom might be less motivated to 'invest' in these matters) pay appropriate attention to them.

Conflict
In a highly differentiated process such as the construction of a building there is considerable potential for conflict between the parties involved. This is often because their objectives are rooted in the organisations they represent, rather than in the project.

4.3 Risk and value management of the on-site process

Having considered the values and risks that arise from the way the participants in the on-site delivery process are organised, we shall now examine risk and value management using the following classification, based on the appropriate issues identified in section 1.2.2 of Chapter 1:

- technical issues
- organisational issues
- operational issues
- financial issues.

The coverage is not intended to be exhaustive, nor is it exclusive. For example, matters relating to design, to health and safety and to the environment are not restricted to the on-site process, though they are dealt with here (under operational issues) as well as elsewhere.

4.3.1 Technical issues

Many of the technical issues are inherited from the building's design. The value of a good design solution lies partly in its response to the technical risks posed by the physical concerns such as geology and climate. The aim of the design solution is to add or maintain *value*. But there are often risks.

Innovation
The construction industry is relatively slow in adopting new technologies, and usually prefers known approaches to innovative ones.

Example

An innovative design can add value in a number of ways: it may be for its own sake; it may be to cope with complexity; or it may be to produce technical efficiency in the materials used, the loads they impose, or the structural feats they can achieve. But innovation brings risks. *Building* magazine[5] reported on the Thames Millennium Bridge that:

It had taken 18 months to develop and install the remedial dampers. The relief was palpable when 4,000 feet failed to produce more than a barely perceptible quiver. The problem manifested itself spectacularly in the summer of 2000, when the Millennium Bridge – a work of pure engineering presented as a work of art – wobbled on its opening day.

4.3.2 Organisational issues

Preparation
Advance planning and organisation are valuable. Time and money spent on organisation is often well spent, but there are two major threats: first, the expenditure could be excessive when compared with what it achieves; second, the correctness of the planning and organisation may be compromised by uncertainty. Despite the obvious risks involved in planning activities (such as contractor's cost estimating, planning, site mobilisation), they are usually carried out without proper assessment of the risks. Traditionally, such risk analysis as there is relies on rule of thumb and the use of contingency sums or percentage additions.

Information and its communication
The availability of information and its proper communication are constant preoccupations. Design information is a particular problem. Design is a complex process, but fees and times for design are constantly under pressure, which can result in extreme risks being put on the timing and quality of what is produced. It is sometimes inevitable that the design information has to be changed. But when change is made to enhance value, the concurrent risks imposed, particularly on the time and cost of the project, should not be underestimated.

4.3.3 Operational issues

Under this heading we have grouped the general managerial aim of improved efficiency together with statutory fulfilment such as compliance with health and safety and environmental legislation.

Reducing costs and increasing value
At the heart of the drive to improve construction is the quest for increased efficiency. Probably the most obvious way to achieve this is to reduce cost, while maintaining the other requirements of the project. A typical approach is that of value engineering. As discussed earlier in Chapter 3 and again later in Chapter 5, *value engineering* (as distinguished from the broader concept of *value management*) is concerned with maintaining (or increasing) *value* while reducing (or maintaining) *cost*. This is best described by the formula

$$V = F/C$$

where V = value, F = function, and C = cost.

So value can be increased in three ways:

- by keeping function constant and decreasing cost
- by increasing function while keeping cost constant
- by both increasing function and decreasing cost.

This type of approach is not without risks. For example, the process may well involve change, and must therefore take into account the disruptive costs of change, which increase as the project proceeds.

Increased demands on time
In the context of completion time these attempts at value engineering seek to improve delivery times for buildings by combining techniques such as concurrent design, parallel working, planned programme compression, and just-in-time delivery. But an unrealistic programme is as great a threat to project success as an unrealistic budget. Not only is time compression of the programme a costly activity, but examples abound where time cut from front-end organisational activities has resulted in longer periods of production and even longer overall times.

Health and safety
Before the Health and Safety at Work Act 1974, safety legislation was essentially prescriptive. The 1974 Act represented a radical change in approach, requiring (among other things) that all employers or other responsible persons do everything reasonably practicable to provide a safe work environment. Subsequent legislation has followed the same approach. For example, the Management of Health and Safety Regulations 1999 require all employers and responsible persons to carry out suitable and sufficient risk assessments, and to act on these. All workplaces with five or more employees must document these assessments. The risk assessment process involves:

- examining the work environment and the activities contained therein to consider the likelihood and severity of harm to people that may be caused
- specifying control measures that would minimise the risk as far as is reasonably practicable.

Environmental
There is a trend to stricter environmental legislation: for example the Town & Country (Assessment of Environmental Effects) Regulations 1988, the Environment Act 1995, and the Environmental Protection Act 1990 (the EPA). The EPA requires an overall risk-based approach to dealing with contaminated sites, which is consistent with the general good practice approach to managing land contamination.

4.3.4 Financial issues

Relevant aspects of risk and value here include the *value* (on the one side) and *risks* (on the other) of protections against contractual default, determination and insolvency, such as *bonds, guarantees* and *warranties*.

Insurance
There are various insurances that the parties involved in a building project are required or would be advised to take out. They range from the statutory requirement to cover the risks of employing people, to the prudent measures taken to cover against injury to third parties or their property. There will be a need to guard against the risk of the building itself being damaged (insurance of the works), and there are also means of insurance (for example professional indemnity insurance) for professionals who design or advise for a fee. Finally, there may be a network of cross-indemnities through which parties relieve one another of the risk of claims by others.

Bonds
A bond is a legal agreement between three parties:

- the *guarantor*, usually a bank or a surety
- the *guaranteed*, usually (though not always) a contractor
- the *beneficiary*, usually (though not always) an employer.

Bonds are used for various purposes, including discouraging bidders from withdrawing their bids (*tender* bonds); encouraging the performance of obligations (*performance* bonds); and other, less common varieties such as *advance payment* and *retention* bonds. There are various types of bond in use, but it important to distinguish between two main types:

- *on-default* bonds (also called *conditional* or *documentary* bonds)
- *on-demand* bonds, for which little or no evidence of default is required. The bond may be called in 'on first written demand'.

4.4   Conclusions

The on-site teams that produce buildings both benefit from and suffer from the fact that they are temporary coalitions. When these parties interact successfully they can enhance value and sustain it throughout the various stages of team development, but the risks of failure are considerable.

The following are the key points covered in this chapter:

- The fragmentation of the on-site process actually adds value, but brings risks.
- Examples of value are the technical and economic advantages of subcontracting.
- Example of risks are the increased difficulties of control, communication and coordination.
- Risk and value are inherent in the site process, and these can be classified as
  - technical issues
  - organisational issues
  - operational issues
  - financial issues.

Notes and references

1 Gray and Flanagan found that in many cases over 90% of the value of projects they looked at was subcontracted. See C. Gray and R. Flanagan *The Changing Role of Specialist and Trade Contractors* (Chartered Institute of Building, Ascot, 1989).

2 The entitlement to an extension of time in such circumstances is not a 'cast-iron' certainty, as seen in the decision in *Westminster CC* v *Jarvis & Sons Ltd* [1969] All ER 942.

3 The argument for this is most convincingly presented in M. Ball *Rebuilding Construction: Economic Change in the British Construction Industry* (Routledge, London, 1988).

4 Use of this as an indicator of a change in approach by main contractors was suggested by Ball, *Rebuilding Construction*, p 94.

5 News feature, *Building* 15 February 2002, pp 22–24.

5 Value management in practice

5.1 Introduction

It is the way a building fulfils its functions, and the efficiency with which it does it, that give it its value. Hospitals are built as places to treat patients, schools to provide education, and offices to work in. Were it not for such things as the need for shelter and security these things could be done in the open air, and there would be no need for a building. The bricks and mortar are simply a means to an end; they do not provide any value in themselves. Although buildings are usually valued in accounting terms, as assets their value is generally not inherent; it is derived from their function. If the need that a building was constructed to fulfil disappears, then the building's value will derive from its ability to be adapted to an alternative use. If it cannot be adapted for another use then the only value will come from the land it occupies (less the cost of demolishing the building – a negative value).

In designing and making buildings it is easy for this to be forgotten, or for the functional issues that add value to be overlooked in rushing to solve the technical and contractual issues. This is not surprising; buildings take a long time to make, and making them involves a large number of people. Also, each building is usually a 'prototype', with a unique owner, client and user: to further complicate things these are usually not the same person. Given the length of time, the number of people and the unique characteristics of each project, it is not surprising that sight can be lost of the original objectives and purpose of the building – the very things that give it its value.

Value management is concerned with ensuring that the best value is delivered. In a perfect world this would, by design, take care of itself. In truth, as with most human endeavours, it requires careful management to ensure and maximise the delivery of value. Given the timescales involved, and the complexity of the building process, it would be very surprising if this were not so.

5.1.1 The background to value management

The roots of value management are normally traced back to the US General Electric Company. In the 1940s Lawrence Miles, an engineer employed by GEC, developed a technique that he called *value analysis*. This involved focusing on the function that a product performed and finding an alternative way of performing this function, as opposed to finding an alternative product. This is fundamentally what value management is about. It is not about substituting machines, or using alternative products, or eliminating functions, but rather about analysing what it is you wish to do and finding the most efficient way of doing it – the way that adds most value.

Example

From time to time you need to repoint the chimney of your house. To do so you need safe access to the roof and a platform to work from. The function you wish to perform is safely repointing the chimney; if you think about it you can imagine a number of ways of doing this. For example you could:

- erect scaffolding
- have a permanent access ladder fixed externally
- extend your staircase to roof level with an access hatch and permanent platform
- hire a helicopter to hover above, suspending you from a harness
- put up a temporary ladder with a man-safe harness
- avoid the function altogether – clad the chimney in something that is maintenance free
- sell the house and move before the chimney requires repointing and so on.

The point is that, by analysing the function, the focus is on *achieving* the end rather than the *means* of achieving the end. Some solutions will be impractical, or too expensive or disruptive – some involve built solutions and some don't. By analysing the function the need can be considered without presuming that the solution is necessarily some variation on a ladder and platform.

In the UK value management techniques were first employed in the manufacturing industries; they were not taken up by the construction industry until the 1980s. (This was an early example of learning from other industries – something that was to become one of the key recommendations of the 1999 Egan Report *Rethinking Construction*).[1] British Airports Authority – the company that Sir John Egan, the author of the report, was head of at that time – employed, and continues to employ, value management techniques on many of its major projects, including for example the Heathrow Express rail link.

5.1.2 Value management, value engineering or value analysis?

There is often confusion over the exact meaning of value management, and how it is distinct from value engineering and value analysis. In truth the distinctions are often very loose. The *Penguin Management Handbook*,[2] for instance, makes no reference to value management, but describes value analysis in some detail.

A commonly accepted distinction is to see the term 'value management' as describing a range of management techniques for analysing and managing the delivery of value in a project. 'Value engineering' is often taken to mean the study of completed designs, and 'value analysis' to mean the study of the completed building. Value engineering and value analysis are therefore component parts of value management.

Value management has been described[3] as

> a structured approach to defining what value means to a client in meeting a perceived need, by clearly defining and agreeing project objectives and how they can be achieved.

The important points to note from this definition are:

- It is a *structured approach:* there are established procedures that follow a set of rules.
- It sets out to define *what value means to a client:* this acknowledges that value is personal to the client and to their perceived need. There are no absolutes; what constitutes value in one situation can be redundant over-design in another.
- The process is concerned with *defining* and *achieving the project objectives* as distinct from establishing the detailed design of the building.

A distinction should also be drawn between formal value management and the informal design process that takes place on all projects. This informal approach involves both conscious and instinctive consideration of alternatives and the rejection or acceptance of these by the individual designer. There is a strong case for suggesting that all designers subconsciously go through this process, and that this subconscious approach derives largely from experience and training. This exploring, testing, accepting and rejecting of solutions can be seen as the very essence of what designers do – the essence of the architect's and engineer's design skill.

However, there are very real limitations to this approach, and in practice it is often not enough to rely on this alone. These limitations result from imperfections in the design process. As discussed in Chapter 3, building design rarely takes place in a perfect way. Frequent problems include the following:

- It is difficult for the newly assembled design team to arrive at a good working relationship with well-developed communication systems. This problem is most acute at the outset of the project, which is the time when many of the key strategic decisions are made and the 'die is cast' both for the building and for the project team.

- The fragmentation of the design and construction functions, and the sequential assembly of the project team, mean that those involved in putting the building together on site (and therefore those with the knowledge of how the designs will be translated into buildings) are not on the team at this time.
- There is generally a high degree of imperfect knowledge – information regarding such things as the ground conditions, the concerns of the planning authority, the ability of the budget allocated to pay for the building, or the availability of special components and materials.
- There are often many preconceptions regarding the building form and design carried by the design team based on their experience gained on previous projects. There may be an inherent resistance to questioning these preconceptions.
- The internal environment (the project) and the external environment are both constantly changing. Changes in fiscal policy, for instance, can mean that what was once good value no longer represents the best solution. For example, the introduction by the UK government of a landfill tax encouraged a move to design solutions that found ways of reusing excavated material on site rather than disposing of it in landfill sites. Such changes can in turn have an impact on the building design.
- Changes made to a part of the project may well have an impact on the general design, which means that the original solution is no longer valid. For example, a decision may be made to add façade sun shading, which in turn means that a high level of air conditioning may no longer be necessary.

Value management can provide an opportunity to stop the project momentarily and step back and examine the validity of earlier decisions. This questioning can take into account changes since the project began, the direction the design has taken to date, and the availability of new information. The involvement of people from outside the project team – even if only the value management facilitator – can also provide a fresh perspective on the project.

5.2   Brainstorming and functional analysis

Two important techniques employed in value management are *brainstorming* and *functional analysis*. Before looking at the techniques of value management itself, it is worth briefly examining what these involve.

5.2.1   Brainstorming

There is a strong tendency for creative thinking both to become mechanised and to be inhibited by the limits of previous individual experience and knowledge.

Brainstorming is a technique to overcome this. It works by assembling a number of people, and therefore drawing on the creative content of a number of minds rather than on one individual mind. To be successful the process must operate within some rules. In particular there must be no restraint on the ideas generated; they can be as outlandish as people want. For this to work a related rule is that there must be no criticism of the ideas. Brainstorming is about generating ideas, not about generating solutions – that comes later. The outlandish ideas are an important step on the way to the solutions.

It requires some skill on the part of the facilitator to create an environment in which this can take place. Some people have a personality that does not naturally embrace this way of thinking, and within building design teams some disciplines often find it easier than others. This way of working and thinking is generally more familiar, for example, to architects than it is to quantity surveyors. It is important therefore that the brainstorming session is not dominated by those most comfortable and most familiar with this way of working. For the process to be successful it is important that a broad cross-section of ideas are put forward.

At this stage the measure of success is simply the number and the quality of ideas that come forward. It will only be when these are examined more closely at a later stage that the true value of the ideas to the project will become apparent.[4]

Example

The client, a retailer involved in selling home improvement (DIY) products, has acquired a number of sites across the UK for the development of new stores. The speed with which the units can be constructed is in many ways more important than the cost of the buildings. (It is of course in effect the same thing: the sooner the building is completed and begins to trade the sooner the store's income stream will begin.)

A group of people are gathered together in a hotel to brainstorm ideas for designing and constructing the buildings more quickly but without greatly increasing the cost. The background, qualifications and experience of the nine people making up the brainstorm group is interesting:

- the facilitator, a former construction company director who has set up a construction consultancy specialising in value management
- a planning manager from a contractor with experience of fast-track industrial buildings
- an architect with retail design experience
- the design director of a steel fabrication company

- the technical manager of a company that manufactures and erects composite wall and roof cladding systems
- a quantity surveyor
- the client's design director
- the client's maintenance manager
- the client's marketing director.

The group meet, many for the first time, the evening before the brainstorming session. The evening meal is used to allow people to get to know each other, but care is taken not to anticipate the actual brainstorming session.

The brainstorming session begins with the facilitator outlining the purpose of the session, together with an explanation of how brainstorming works – the 'rules of engagement'. He stresses that there is no such thing as a bad idea in a brainstorming session, and that in many ways quantity will be as important as quality. The wall is clear, and the facilitator has a large supply of yellow Post-it® notes on which to write ideas and stick them to the wall. The session is split into two. The ideas-generating session (about two thirds of the time) comes before coffee; after coffee the balance of the time is given over to recording and describing the ideas generated.

The tendency at first is for people to suggest ideas from the perspective of their own experience and training, and to come up with ideas that move only a little way from the present design. The marketing director and the architect tend to dominate proceedings at first, although the facilitator is careful to draw everybody into the discussion. Techniques he employs to do this include:

- taking the idea of one person and linking it to that of another more reticent participant
- calling on people by name to contribute
- politely cutting short the more verbose participants and allowing others more time to develop and express their thoughts
- moving about the room and standing next to the person speaking (this encourages people not used to addressing large groups to feel the atmosphere as conversational)
- maintaining eye contact, smiling, and nodding encouragement to the person speaking.

Care is taken not to criticise or pass judgement on ideas. The facilitator skilfully steers the discussion away from this. The pace is brisk, and once the idea is understood the discussion is moved on. The facilitator is a good communicator, who recognises the importance of non-verbal communication. He is also careful to remain neutral and not bring his own agenda to the discussion. Although he is a 'construction person' this experience is used only to assist understanding rather than to contribute a view.

5.2.2 Functional analysis

Functional analysis is the analysis of objects by their function (*what they do*) rather than their form (*what they are*). As an example, the functions of an external glazed door can include:

- providing security
- keeping rain out
- providing insulation
- allowing access and egress
- allowing daylight in
- allowing views out (and in).

A functional analysis of the door can allow the following:

- Consideration of alternatives that could equally provide these functions.
- A breakdown of the functional costs (and therefore the ability to analyse the value of these). The proportion of the total cost of the door attributed to each individual function can be examined in isolation. For example, the costs of the glazing in the door – forming the opening, the glass, framing this and so on – are all incurred to provide the functions of letting daylight in and allowing views out. This is a very simple example, but attributing costs to individual functions is rarely an exact science; it generally involves a strong element of judgement.
- A critical analysis of how necessary some or all of the functions are in relation to their cost.

Examining a function in this way can allow the creation of a hierarchy of primary or necessary functions, and of secondary or desirable functions.

This technique can inform decision-making in value management. However, in most construction projects a detailed functional analysis and costing of individual elements of a building would probably not be cost-effective. This is why detailed functional analysis is rarely carried out in construction project value management. There are by contrast clear benefits for manufacturing industries that have repetitive production runs, in analysing function and costing this in some detail. An exception in the construction industry might be volume house builders, who may well find that the return justifies the cost.[5]

5.3   Techniques and methodologies

Although value management should be a constant process throughout the project, it is often advisable, certainly on large complex projects, to have at least one formal value management workshop using a professional, experienced facilitator. To be successful the workshop should involve a representative sample of the project team, and should include all the key disciplines and – importantly – the client. It is generally most successful when the client body is represented both by those who will *use or operate* the facility and by those who will be responsible for *maintaining and managing* the building. The essential principle is that the workshop should involve as wide a spread of project stakeholders as possible, without becoming too unwieldy to manage.

There is a well-established format for a 40-hour workshop, although there is in theory no time limit to the exercise.[6] In practice this will be dictated by the available resources and by the needs of the project programme. The cost of the value management workshop should also be carefully considered. Often the most senior members of the project team will attend the workshop, and the high costs resulting from this should not be overlooked. (The workshop should not be attended exclusively by senior project members; it is important that it benefits from the expertise of those involved in actually designing and actually building.)

For the value management process to work successfully it must be carefully structured. Normally the workshop or workshops follow a sequence known as the *job plan*. This comprises five distinct stages, which are examined below. It is usual for some basic design work to have been carried out before the workshops are embarked on, as this will form the focus of the analysis.

5.3.1   Stage 1: Information phase

At this stage the details of the project and the project objectives are presented to the workshop participants. Often the principal designer (usually the architect) or the client will make a presentation describing why the design has developed the way it has; the key functions that the building is to fulfil; the background to the project; and any special conditions that might apply. This presentation should be as objective as possible. Value judgements or personal preferences should be avoided, and the language of the presentation should be neutral and descriptive.

A formal functional analysis can also be carried out at this stage, although the cost-effectiveness of this will need to be carefully considered.

A further important facet of the information phase is to allow the participants to get to know each other. As the workshops will involve challenging the work, preconceptions and preferences of others, a good working relationship must be fostered as early as possible. For this to be positive it is essential that the facilitator has good people management skills, including:

- *leadership* – the ability to lead the session and allow ideas to develop while ensuring that there is some direction and structure to the session
- *listening skills* – the ability to pick up on ideas that may be difficult for the participant to articulate (it is often the best ideas that are most difficult to express, either because of their complexity or because of their simplicity)
- *editing skills* – the ability to take raw ideas and set these down as a potential plan of action for future development
- *meeting-management skills* – the ability to encourage input from all the participants, not only those who would dominate the proceedings in the absence of some control.

5.3.2 Stage 2: Creative phase

This is essentially the brainstorming session when ideas are generated and explored. All ideas must be treated as possible, no matter how outrageous they may seem. The objective is not to generate solutions but rather to come up with ideas that may merit further exploration at a later stage. It is therefore important that no editing or rejection takes place at this stage, and that all ideas are recorded in their raw form for later evaluation.

5.3.3 Stage 3: Evaluation or judgement phase

In this phase the ideas generated are put forward, and their merits are assessed by the workshop participants as a group. Some – probably the majority – will be rejected; this is not a problem. Ideas that are rejected may include:

- those that have already been rejected in earlier design consideration by individual team members
- those that would take time beyond the programme constraints to explore and develop
- those for which the likelihood of a working solution emerging is too much in question to justify the time spent on evaluation and development
- those that are uneconomic (that is, those that would not deliver sufficient added value)

- those that are impractical. If the session has been productive there should be plenty of these. Too few probably means that not enough radical ideas have been explored.

Some of the ideas will be carried forward, and it is often helpful if an individual — not necessarily the generator of the idea — acts as 'champion' of the idea to be developed. This commitment to an idea can ensure that it is properly explored, and not abandoned by default.

5.3.4 Stage 4: Development phase

Further work is carried out at this stage to decide whether or not an idea should become a change to the design or the brief. In effect a mini feasibility study is performed on each idea to see whether it is a viable proposal. This may involve members of the workshop working across disciplines to examine the idea thoroughly. As this may require detailed design consideration it will usually make sense to do part of this outside the workshop. It is important to strike a balance between carrying out a complete redesign exercise and merely doing sufficient work to be satisfied that the proposal actually is a runner (that is, that you are not kidding yourself). It is important to take a holistic view of this. Often solving one problem can create another; or, for example, capital costs may be saved but at the expense of the long-term running costs or the efficient operation of the finished building. Some proposals may appear to improve the finished product but may adversely affect the construction. For instance, the lead-in time for the materials required to construct the building in an alternative way may cause unacceptable delays to the overall programme.

5.3.5 Stage 5: Recommendation or presentation phase

At this stage the ideas are presented to the client. All the costs, consequential costs, benefits and disbenefits must be comprehensively considered. The impact on the function of the building is the key factor. The meeting should contain a cross-section of the client organisation — those who will use the building, those who will own and maintain it, and those who will pay for it. The client team should also include the decision-makers who can authorise the change. Once the client's authority has been obtained the necessary changes to the design can be implemented.

Example: the trail of an idea – structural roof trays
A large retailer set out to explore ways of producing their buildings more quickly. Although the cost of the building was a concern, they were prepared to

pay a premium for a faster construction programme in order to open for trading earlier. The logic of this is obvious when one considers that the weekly turnover of a food superstore can frequently exceed £1,000,000. To explore this a value management workshop was set up. The store layout and the elevations and section had already been agreed, and planning permission had been received. The workshop involved a cross-section of the client organisation, the design and construction team, and importantly the key subcontractors and suppliers.

One key way to improve the construction programme was to design the building so that it could be made watertight sooner. This would have a significant effect on following trades, in particular in laying the floor, and therefore on the critical path. The roof construction consisted of steel beams on a 6 m grid spanning a clear space of 25 m. Purlins spanned between the beams; the roof covering consisted of a pressed metal tray fixed to the purlins containing the roof insulation, on top of which was fixed a standing seam metal roof.

During the creative phase the idea was put forward of eliminating the purlins and using a structural roof tray capable of spanning the 6 m between the roof beams. However, during the evaluation phase it became apparent that the open zone created by the purlins between the roof deck and the top of the beams was not simply 'left-over space', but fulfilled an important function as a space to run lighting cable trays and the fire sprinkler pipe system. Although this was an important concern, it was nevertheless considered that the benefits created would outweigh this problem, and it was decided to take the idea forward to the development phase and explore ways of overcoming this problem.

During the development phase an alternative route for the cabling and pipework was developed. Costings of the original and the revised design were prepared; the aesthetic impact was tested by producing three-dimensional visualisations of the two alternatives for comparison; and finally the programme implications and benefits were fed into the master programme. A report was prepared containing this information and a presentation made to the client (which included the operation and maintenance staff). Not only did the client accept the change, but they incorporated this design into their store standard design manual for future use.

5.4   Leading and participating

There are numerous construction and management consultants who facilitate and run value management workshops, but there are very few people who make a living exclusively from value management. Hence most people in the construction industry are likely to come across value management when they participate in a formal value management exercise.

However, there is no reason why value management cannot be applied as part of the normal design process on most projects. Because of the size, complexity and number of participants, construction projects have a tendency to develop a life and direction of their own. There are many potential benefits in briefly halting the design development and questioning the direction that the design is taking. Applying the steps of informing, creating alternative ideas, evaluating, developing and implementing gives this a rational and productive framework. The use of brainstorming and functional analysis, involving a well-selected cross-section of the project and client team, can yield ideas that add additional value to the project. The process can also often contribute positively to team building, motivation and the flow of information, all of which will provide further value.

In many ways the process of value management complements the way in which designers think. Much of what is an explicit procedure in value management is what good designers do subconsciously. The added benefit is the ability to draw on the minds (the knowledge and experience) of others, and to apply some objective quantification and value judgement to the ideas generated.

5.5   Value management in context

Architects often perceive value management as a threat both to the architectural profession and to the quality of architecture. These concerns are born of a number of factors, many of which arise from a misunderstanding of the nature of value management.

There is a concern that the entire process is redundant – that the consideration of alternative design solutions is a function carried out by all designers, and so there is no need for any outside involvement in this process. This ignores the design management problems inherent in construction projects, which mean that the design is rarely carried out in the optimum way. It also ignores the reality of the commercial pressure that the design team is subject to. The ability to take time to step back and 'cross-examine' the design solution can potentially yield vast benefits. The role of the facilitator is not to usurp the design team's role but to *facilitate* the search for value by creating and managing the formal framework within which this search happens.

Concerns that the quality of architecture will be diminished are often based on a misunderstanding of the difference between cost and value. Value management is not about cost cutting – although it may well lead to savings in cost. It is not about reducing the specification for the project; it is about adding value to the project. This view also arises from a narrow interpretation of architecture that emphasises the form over the function of the building. In great buildings the two are complementary. Indeed it can be argued that a sophisticated view of value will recognise the benefits that derive from outstanding buildings.

Example: how value arises in buildings

The value inherent in a good building can create benefits or added value that extend beyond the building itself and beyond its primary function Some of these benefits can be quite unexpected, both in their nature and in their magnitude. The Guggenheim Gallery, Bilbao, by the American architect Frank O. Gehry has given rise to something called the 'Bilbao effect' . Not only has the building increased tourism to what was once a run-down industrial city, there is also strong evidence that it has significantly contributed to the economic regeneration of the Basque region.

The inherent value arising from good design can extend the life of a building. Bankside Power Station on the South Bank of the Thames was considered such a significant building that, when its functional life ended, rather than demolishing it a new use as an art gallery was found for the structure.

These examples illustrate how the value of good design can be difficult to define but can have important economic implications.

Value management cannot be considered in isolation. For the construction industry it is a project management tool that sits alongside risk management, quality management and cost management as part of the total project management system. It is probably not appropriate or cost-effective for small projects, although the techniques of brainstorming and functional analysis are useful at some level in many project situations.

Value management is complementary to and shares many of the concepts that have developed in concurrent engineering and lean construction.[7] *Concurrent engineering* is concerned with moving away from 'over the wall' design. It promotes the idea of designers going forward and developing the design together, rather than the traditional model by which a designer develops a design that is thrown over the wall to the quantity surveyor, who costs it and throws it back because it exceeds the budget. When this process is repeated with contractors, engineers, and so on, and when it happens several times in a project, inevitably cost and time are wasted.

Lean construction is concerned with analysing the processes involved in design and construction, and seeking out and eliminating procedures and practices that are inefficient or that don't add value.

There is an argument that value management is of most benefit where large quantities of a component are to be produced, because even small savings in manufacturing costs can result in significant savings when multiplied across the production run. Although this is the way things are in manufacturing, it is not – with the exception of prefabrication and modularisation – the way things usually are in the construction industry. However, although more benefits may accrue in manufacturing, there will still be some benefits in construction. As with any management technique, the important point is that the time spent should be cost-effective: that is, the benefits arising – improvements in function, reduction in cost, and so on – should exceed the cost of the exercise. This is something that will require careful review before embarking on a value management exercise.

It is also true that on projects that are part of a long-term strategic partnering alliance, the benefits derived from value management can be recovered across a number of projects. This is dealt with more fully in Chapter 9.

The case for value management can be made on three fronts:

- It is cost-effective. Anecdotal evidence suggests that a value management study carried out for a fee of 1% of the construction costs can generate cost savings of the order of 10–15%.
- It improves both the briefing and the design process. The involvement of a multidisciplinary team examining the brief and interrogating the design will inevitably yield improvements.
- It finds the unnecessary project costs. Traditional cost planning makes savings to reduce cost; value management finds the unnecessary costs. At the end of the day the cost of the building may be the same, but the value will be higher.

The savings arising from eliminating unnecessary construction costs can either be taken as savings or spent elsewhere on the project where they actually add value. In this way value management can be seen to be a force for good architecture.

5.6   Conclusions

The techniques of value management include:

- *brainstorming* – an established technique for obtaining the ideas of a large number of people
- *functional analysis* – a technique for analysing a component by its function, that is, by what it does rather than what it is.

The five stages that a value management workshop follows are:

- the *information phase* – presentation of the project objectives and the design work completed so far
- the *creative phase* – brainstorming and searching for imaginative alternative ways of fulfilling the brief
- the *evaluation phase* – testing these ideas
- the *development phase* – taking forward the ideas that are feasible and will increase value
- the *ideas and presentation stage* – presenting the developed ideas, costed, and with their benefits defined. The intention is to seek approval for change.

The participants in a value management workshop should include a broad cross-section of the design, construction and client team. The client team should in particular include those who will operate, finance, manage and maintain the facility.

Value management is unlikely to be cost-effective for smaller projects, although many of the techniques can be applied in an abbreviated form and will assist in brief building and design development.

Value management can provide the following benefits:

- It can generate savings of 10–15%.
- It can improve the briefing and design process.
- It can find the unnecessary project costs.

Notes and references

1 J. Egan, *Rethinking Construction: Report of the Construction Task Force to the Deputy Prime Minister* (Department of the Environment, Transport and the Regions, London, 1998).
2 T. Kempner (ed.) *The Penguin Management Handbook* (Penguin Business, 1987).
3 J.N. Connaughton and S.D. Green *Value Management: A Client's Guide,* CIRIA special publication 129 (Construction Industry Reasearch and Information Association, London, 1996).
4 For further reading on brainstorming and the thought process in design see B. Lawson *How Designers Think,* 2nd edn (Butterworth Architecture, London, 1990).
5 For a more detailed example of functional analysis see the excellent A. Ashworth and K. Hogg *Added Value in Design and Construction* (Pearson Education, Harlow, 2000).
6 For further details of how a 40-hour workshop can be structured see J. Kelly and R. Poynter-Brown 'Value management' in P. Brandon (ed.) *Quantity Surveying Techniques: New Directions,* pp 54–64 (Blackwell Scientific, Oxford, 1990, reissued 1992).
7 For a further insight into lean construction see G. Howell and G. Ballard 'Lean production theory: moving beyond can do' in *Proceedings of the Conference on Lean Construction,* Santiago, Chile, September 1994.

6 Risk management in practice

6.1 Introduction

In its broadest sense, there has always been risk management in construction activity. There is no doubt that the industry contains its fair share of risks, be they technical, physical, commercial or environmental. Construction risks have always been recognised, and the more prudent sponsors of building work and their managers have always made allowances for them. Formal management of risk using scientific approaches is relatively new, but it has become part of every aspect of construction activity and is even (in the case of health and safety) enshrined in the language of legislation.

Although formal risk management in construction is relatively recent, the techniques that it employs have a long pedigree. With very few exceptions, the mathematics of risk, and in particular theories of probability, were all developed in the hundred years between 1650 and 1750.[1] At first, the application of these advances was limited to puzzle-solving and gambling. But there followed a period of massive change in the way society, and in particular business and trade, was organised; probability theory became indispensable to decision-making in some areas of business, especially in the insurance sector.

If the seventeenth century provided the mathematical tools, and the eighteenth century gave them a serious role in society, it was in the nineteenth century that thinkers[2] began to expound the principle that problems of uncertainty could be dealt with by using probabilities derived from observation and past experience. The twentieth century saw the earlier mathematical tools used in the development of game theory, first in the hands of individuals[3] and later, once its military and business significance had been identified, by the think-tanks of powerful organisations such as the Rand Corporation in the USA. Thus the stage was set for the modern era of risk management.[4]

6.2 Definitions, approaches and some basic theory

Before considering methods of risk management and the techniques they adopt, it is worth looking briefly at the word *risk* and what it relates to. This is followed by an introduction to *expected value*, a concept that underpins all risk management theory.

6.2.1 Risk and uncertainty

The fundamental thing about risk is its association with the unknown. In common usage, the concept often has to be disentangled from that of *uncertainty*. One way of differentiating them is the following observation:

> *Risk is measurable uncertainty; uncertainty is unmeasurable risk.*

Example

A hotelier is thinking about building a new extension, and is wondering how much it will cost. He takes advice from a quantity surveyor, who provides an estimate. The hotelier asks whether he is certain. The quantity surveyor (of course) replies with the answer 'no' and gives two more estimates, one higher than the original one and one lower. The hotelier asks which end of the range is more likely. The quantity surveyor replies that the market is 'quiet', so tenders might come in towards the lower price. This typical dialogue shows how the situation has advanced by degrees from one of complete uncertainty to one where a risk assessment (and subsequently a decision) can be made.

6.2.2 Risk and hazard

Hazard is another word that is commonly associated with risk. Again, the meanings of the two words should be kept distinct. Hazards are pre-existing conditions that have the potential to inflict some negative impact: the concept of risk takes account of other circumstances in order to evaluate this potential. An illustration of this distinction can be seen in the following example of methods of travel.

Example

The hazards of motorway driving are inherently greater than those of driving on minor roads, owing to the speeds encountered. However, statistics show that the risks are relatively lower. The risks are kept low by the design of the system: for example, the motorway layout ensures that traffic moving in opposite directions is separated by a central reservation and by barriers.

6.2.3 Risk as probability

Note the role in the last example of statistical evidence to demonstrate risk. It is now common to see risk expressed as the *probability*, or likelihood, of a particular event's occurrence.

Example

Today's weather forecast gives a 70% chance of rain in the north-west. Such a forecast would normally be enough to make many people take precautions in the form of a raincoat or umbrella.

6.2.4 Risk as impact

Another way of looking at uncertain events, especially if it is desirable to prioritise them or choose between them, is in terms of what *impact* they would have.

Example

A businessmen is faced with the question of whether or not to take a particular course of action. If this action pays off, there will be a £20,000 gain; if it fails, there will be a £50,000 loss. Given the information, the gamble really doesn't seem worth it.

6.2.5 Risk as impact times probability

A breakthrough in the analysis of risk, and the concept that underpins all risk management theory, comes when the two approaches are combined. This is evident in the definition of risk provided in British Standard 4778:

Risk is a combination of the probability, or frequency, of the occurrence of a defined hazard and the magnitude of the consequences of the occurrence.

This approach to risk is not a new one,[5] but what distinguishes it is that both the *impact* and the *probability* of the risk are taken into account. Consider this in its most simple form, where impacts and probabilities are simply categorised as either *high* or *low*. A risk matrix can then be designed and actions taken according to its results. Figure 6.1 illustrates this.

The appropriate risk responses can then be decided upon. They may be as follows:

- *Situation 1* – Low probability/low impact. This is the least worrying combination. It might be appropriate to ignore the risk under some circumstances.
- *Situation 2* – High probability/low impact. These are risks that are often catered for by some form of contingency allowance.
- *Situation 3* – Low probability/high impact. This is the type of risk that is typically dealt with by insuring.

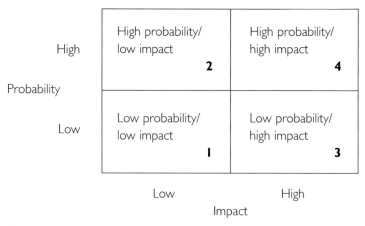

		Low	High
Probability	High	High probability/ low impact **2**	High probability/ high impact **4**
	Low	Low probability/ low impact **1**	Low probability/ high impact **3**

Low — High

Impact

Figure 6.1 Impacts and probabilities

- *Situation 4* – High probability/high impact. These are the risks that would give most concern. The level of concern might be sufficient to warrant avoiding such risks by refusing to take on the project.

6.2.6   Risk and the concept of expected value

'High' and 'low' are useful ways of distinguishing impact and probability, but they are hardly suitable for fine-tuned decision-making. However, when a figure can be put against the impact and the probability of a risk – that is, when they can be *quantified*, either because they are known or, more likely, can be estimated – it is possible to go a step further with the evaluation. The product of multiplying impact and probability is usually referred to as *expected value* (EV). It is the most basic unit of risk measurement.

Returning to the simple example considered earlier, the availability of more data might change the decision.

Example

The same straight decision is required as to whether or not to take a business gamble. There is an 80% probability of a £20,000 gain and a 20% probability of a £50,000 loss. The problem can be formulated as follows:

Win/(Loss)	Probability	EV
£20,000	0.8	£16,000
(£50,000)	0.2	(£10,000)

Should the gamble be made? The expected value (EV) of the alternatives is:

£20,000 × 0.8 = £16,000

plus

(£50,000) × 0.2 = (£10,000)

giving a net probable benefit as a result of the gamble of

£16,000 − £10,000 = £6,000.

On this purely mathematical basis the gamble might appear to be worth taking.

6.3 The process of risk management: emergence of formal methods

An appreciation of the way impacts and probabilities are combined to produce expected values is fundamental to understanding all risk assessment techniques, but before any more sophisticated methods are introduced it is necessary to broaden the picture and look at the management of risk in its entirety. This includes not just the analysis or assessment of risks, but the way they are identified in the first place, and how they can be managed. There have been various attempts to produce standard methodologies for dealing with risks. These include:

- *Risk Analysis and Management of Project Risks* (RAMP), produced jointly by the Institution of Civil Engineers, the Faculty of Actuaries and the Institute of Actuaries
- *Project Risk Analysis and Management* (PRAM), produced by the Association of Project Managers.

Despite certain differences there is a substantial consensus shared by all the standard methods, which tend to be generically applicable across disciplines and industries. The processes they describe invariably involve the following stages:

- identify
- analyse
- evaluate
- allocate
- manage.

These headings will form the basis for the treatment of risk management in this chapter. The next section will discuss ways of *identifying* risks in the first place;

there will then be an introduction to some of the techniques available for *analysing, assessing,* or *evaluating* risks; this is followed by a description of the way risks can be *allocated,* and finally by a section on the *control* or *management* of residual risk – that is, the identified and analysed risks that have not be allocated elsewhere.

6.3.1   Techniques for identifying risk

In order to manage risks it is essential to identify them. Most disasters happen because the potential risks were not envisaged in the first place. There are a number of techniques available for identifying risks. Some risk theorists have categorised these as *intuitive, inductive* and *deductive* techniques, as shown in Table 6.1.

Intuitive techniques
These techniques are generally built around some sort of activity that relates to what is commonly described as *brainstorming.* Brainstorming can be structured or unstructured, but the same general rules usually apply:

- Always evaluate later (don't pause to reject 'silly' ideas).
- Always go for quantity (rather than apparent 'quality').
- Always encourage wild ideas (they might contain 'gems').
- Build on and combine ideas (it might be that the combination of two or more is important).

Table 6.1 Techniques for identifying risks

Category	Basis	Examples
Intuitive	Let's say…	Brainstorming Delphi technique
Inductive	What if…?	Prompt lists HAZOPS FMECA
Deductive	So how…?	Hindsight reviews Risk registers Accident investigation & analysis

Delphi technique
The Delphi technique[6] can form the basis of *group expert brainstorming*. It is designed to pool expertise while removing interpersonal barriers. Typically, a questionnaire is given to members of the group, who are at physically separate locations. Views are taken and distributed. Revised views are collected, and the process is repeated until a consensus ensues, or further iterations are considered unproductive. The distinguishing feature of the technique is that the contributors are kept apart and their contributions are anonymous (as was the Delphic 'voice' that gave consultations to the ancients). The advantage of this approach to pooling opinion is that influences of peer pressure, seniority, or strong personalities are removed. The technique is therefore of use not only in identifying risks, but also in evaluating them.

Inductive techniques
Inductive techniques are those that use expert prediction to identify 'what could go wrong'. The best-known example of such techniques is probably *Hazard and Operability Studies* (HAZOPS). The technique was first used by ICI in its design of chemical plants, and has subsequently been adopted by many other organisations. It begins with a form of structured brainstorming by a multidisciplinary team of experts, who carry out a systematic review of every plant item, noting both its purpose and any possible deviations from the fulfilment of that purpose. The deviations are then investigated in more detail and, where possible, changes are made to eliminate them or to limit their impact.

Another form of inductive technique, *Failure Modes and Effects Criticality Analysis* (FMECA), is normally performed by an individual with expertise in the project in question. The approach involves the diagrammatic representation of systems, their components and possible modes of failure. Whereas experience-based checklists can become hidebound by 'boxed-in' thinking, in which only 'the usual' risks emerge, the use of prompt lists can stimulate risk identification in an 'off-the-wall' way by combining ideas randomly. Table 6.2 shows an example.

The identification of risks: ad hoc or systematic?
Another way of categorising risk identification is to decide whether the process is a specific or ad hoc one (that is, used in isolation and picked up or laid down whenever required), or whether it is more systematic, and reiterated from project to project. Indeed, to be effective, systematic risk identification relies on the codified experience of past project risks.

Advantages and disadvantages of being systematic
There are advantages and disadvantages to being systematic. On the one hand it ensures rigour and consistency, and promotes learning through the use of feedback. In this respect a systematic approach to risk identification seems

Table 6.2 Example of a prompt list

Competence	Ignorance	Competition
Fallibility	Uncertainty	Change
Communication	Site investigation	Non-completion
Transport	Settlement/landslide	Abandonment
Team work	Leakage	Errors
Forecasts	Flooding	Defects
Assumptions	Fire	Practicality
Permissions	Explosion	Terrorism
Instructions	Contamination	Vandalism
Approvals	Pollution	Corruption
Long delivery	Archaeological	Programme
Regulations	Services	Temporary works
Access	Performance	Cost
Confined spaces	Briefs	Legislation
Safety	Security	Royalties
Toxicity	Availability	Patents
Health	Compatibility	Liabilities
Accidents	Reliability	Warranties
Impact	Durability	Tax
Weather	Buildability	Users
Earthquake	Repairability	Labour relations
Theft	Workmanship	Subcontractors
Environment	Commissioning	Public

preferable. On the other hand, being systematic does not guarantee successful risk identification. Failure can occur for a number of reasons:

- Systems have the habit of becoming bureaucratic duties that are observed without regard to what they were designed to achieve.
- Risk blindness may occur because of the contextual bias of the person or group carrying out the exercise. For example, an engineer may be strong on the identification of technical failure but unable to predict failures of communication or management systems.
- Risk identification systems might be self-defeating by definition, in that systems cater for normal situations and the most critical risks lie outside that definition!

Perhaps an effective risk identification system should comprise a base-line systematic approach *and* an element of 'out-of-the-box' thinking.

6.3.2 Techniques for analysing and evaluating risk

Having identified the potential risks, the next stage is to analyse and evaluate them. There are basically three approaches, of increasing complexity:

- epistemic
- stochastic
- aleatoric.

Epistemic approaches
Epistemic methods (from the Greek *episteme*, 'knowledge') are based on understanding, or experience. An example of such a method is *sensitivity analysis*. Imagine a decision involving the output rate of an excavator. There may be two main factors that affect the output:

- the depth of dig
- the type of soil.

Figure 6.2 shows an example of a sensitivity analysis for the excavator. Note the differing sensitivities: changes in the depth of dig (variable 1) will have a relatively small effect on the output of the machine (shown on the x-axis), whereas a relatively small change in the soil type (variable 2) will have a greater effect. Where a number of variable factors exist it may be possible to combine them on the same diagram.

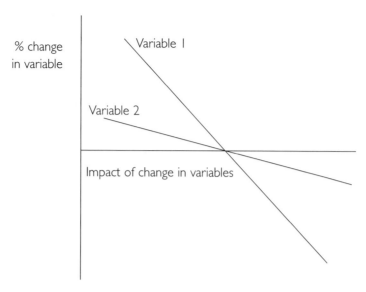

Figure 6.2 Sensitivity analysis: analysis for the excavator

Simple epistemic methods are limited by the fact that the values of all the variables are assumed to be equally likely. There is no attempt to assess different probabilities for different values of the variables.

Stochastic approaches
Stochastic methods, by contrast, incorporate *probability* into the risk analysis.

Common techniques include *risk matrices* and *risk decision trees*. The basis of both techniques is to assess a risk by multiplying its impact by its probability. The simplest example is the risk matrix. These can be qualitative or quantitative. The example given below is based on a health and safety risk analysis carried out on a building design.

> **Example**
> The following is a design risk assessment for health and safety purposes. As usual, the process includes the following stages:
>
> 1 Identify hazards.
> 2 Assess the severity of their impact, and upon whom.
> 3 Assess their likelihood.
> 4 By combining 2 and 3, evaluate the risk.
> 5 Design risk control measures.
> 6 Implement them and review to ensure effectiveness.

Hazards may be identified using a variety of methods or codings. An example is given in Table 6.3, which shows the severity of the impact of particular hazards.

Continuing with the same example, Table 6.4 includes an assessment of the likelihood of the hazards occurring.

Combining the scores in Tables 6.3 and 6.4 gives a resultant so that Table 6.5 can be used to evaluate the risk and decide on appropriate control measures. For example, a hazard that was *serious* (e.g. 4 in Table 6.3) and *possible* (e.g. 4 in Table 6.4) would result in a risk value of 16. According to Table 6.5 this risk merits *reduction* by modification of the design.

Table 6.3 Health and safety risk assessment: severity of the hazard

Severity	Weighting	Description
Minor	1–2	Slight or minor injury
Serious	3–4	Serious to moderate injury
Major	5–6	Loss of life or severe injury

Table 6.4 Health and safety risk assessment: likelihood of the hazard

Likelihood	Weighting	Description
Improbable	1–2	Would be unlikely to occur
Possible	3–4	Could occur
Probable	5–6	Would be likely to occur

Table 6.5 Health and safety risk assessment: evaluation of the risk and control measures

Risk result	Action	Description
1–8	Avoidance	Design out the identified hazard (avoid creating or increasing others)
9–19	Reduction	Modify the design to reduce hazards (avoid creating or increasing others)
20–36	Control	Modify the design to give acceptable safeguards against residual hazard

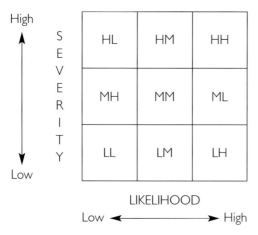

Figure 6.3 Health and safety risk assessment: alternative expression of risk assessment

Figure 6.3 is a further example of a simple expression of risk, using the letters H (= high), M (= medium) and L (= low) to denote both the impact and the probability of a particular hazard. The resulting matrix ranges from very high (HH) to very low (LL) risks.

Risks can also be evaluated using the *decision tree* approach. Here impacts and probabilities are assigned quantitative values, based upon experience and multiplied. The example in Figure 6.4 shows how the techniques can be used to produce a probabilistic time period for a three-activity schedule. Activity A has a 60% (0.6) chance of completion in 8 weeks and a 40% (0.4) chance of a 10-week completion. The probabilistic duration is therefore 8.8 weeks.

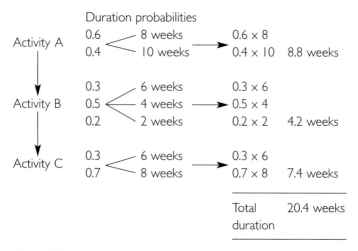

Figure 6.4 Decision tree

Table 6.6 Example of rolling a pair of dice 100 times

Value	Frequency	Cumulative frequency
2	1	1
3	6	7
4	4	11
5	14	25
6	15	40
7	17	57
8	10	67
9	14	81
10	10	91
11	8	99
12	1	100

Aleatoric approaches
Simple stochastic methods are limited by the fact that the values of all the variables are assumed to be independent, and that assessment of the probabilities usually involves a high degree of subjectivity.

Aleatoric (from the Latin *aleator*, 'dice-thrower') methods bring the added power of simulation to the assessment and evaluation of risk. They involve the use of probability distributions. A probability distribution is a way of presenting a sample of occurrences. Take the simple example of rolling a pair of dice 100 times. The results are given in Table 6.6.

Results such as these can be presented in a variety of ways, for example as a *histogram* (Figure 6.5), or as a *probability distribution* (Figure 6.6).

The *shape* of the probability distribution reveals a great deal about the nature of the event in question. For example, the spread (measured statistically as *variance* or *standard deviation*) indicates the level of uncertainty of outcome. Compare Figures 6.7 and 6.8.

The distributions shown are bell-shaped. However, probability distributions can be presented in simpler forms, for example rectangular (Figure 6.9) or triangular.

Rectangular distributions (where there is an equal likelihood of a number of values of the variable) are unlikely to be truly representative, and are not

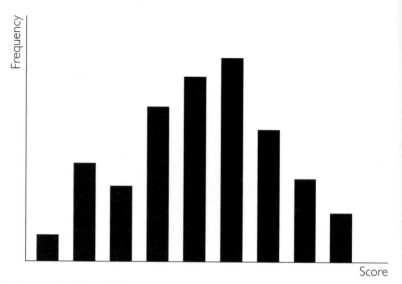

Figure 6.5 Example of rolling a pair of dice 100 times: a histogram

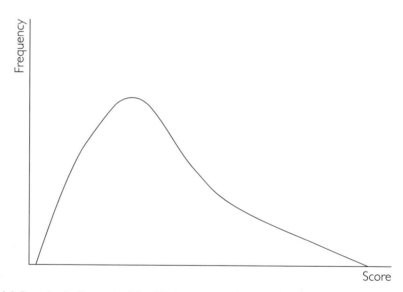

Figure 6.6 Example of rolling a pair of dice 100 times: presented as a probability distribution

particularly useful for our purposes. Triangular distributions, however, though they are more general approximations are relatively realistic and considerably easier to use in calculations. An example of a triangular probability distribution is given in Figure 6.10. This represents (in simpler form) all the aspects of the Gaussian, or bell-shaped, probability distribution, an example of which is given in Figure 6.11.

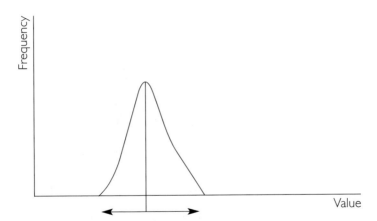

Figure 6.7 Small standard deviation = low uncertainty

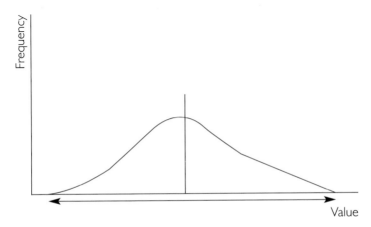

Figure 6.8 Large standard deviation = high uncertainty

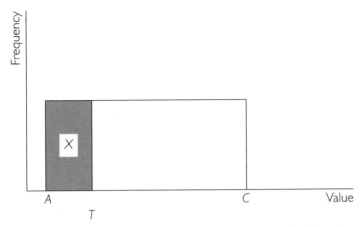

Figure 6.9 Rectangular distribution. A = minimum value; C = maximum value. There is an equal chance of any value in between. X is the probability density for a given value (T)

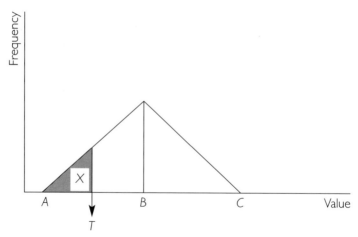

Figure 6.10 Triangular distribution. A = minimum value; C = maximum value; B = most likely value. X is the probability density for a given value (T)

Besides its basic shape, the *skewness* of the probability distribution demonstrates important things about the event in question (Figure 6.12). Unskewed (that is, symmetrical) distributions demonstrate an equal likelihood of high-value and low-value events. When the probability distribution is skewed to the left, it indicates that lower values are more likely; when it is skewed to the right, the opposite is true.

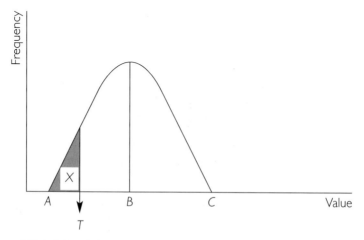

Figure 6.11 Bell-shaped distribution. A = minimum value; C = maximum value; B = most likely value. X is the probability density for a given value (T)

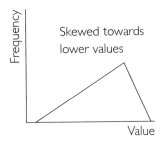

Figure 6.12 Skewness of distribution

Probability distributions can be used to calculate probabilities at given confidence levels. For example, in the triangular distribution illustrated in Figure 6.13 there is 75% confidence that the value of the variable will fall within the shaded area of the probability distribution.

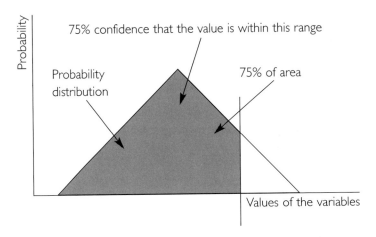

Figure 6.13 Risk simulation

Example

The calculation can be illustrated using a simple example. A project comprises four activities (numbered 10, 20, 30, 40). Their minimum (A), maximum (C) and most likely (B) estimated durations are shown in Table 6.7. Continuing the calculation produces the results listed in Table 6.8.

Table 6.7 Risk simulation: a simple example

Activity number	Duration			Distribution shape	Duration at confidence	
	A	B	C		80%	90%
10	4	8	10	△		
20	5	7	12	△		
30	5	5	8	△		
40	2	4	6	△		

Table 6.8 Risk simulation: solution

Activity number	Duration			Distribution shape	Duration at confidence	
	A	B	C		80%	90%
10	4	8	10	△	8.45	8.9
20	5	7	12	△	9.35	10.1
30	5	5	8	△	6.67	7.05
40	2	4	6	△	4.74	5.37

Probably the most comprehensive and objective approach to modelling risks is to incorporate probability distributions into a *simulation*. If a large number of simulations is required, it is helpful to perform such operations by computer.

Several software packages are available. The principles of their workings are as follows:

1 Select the variable(s) whose effect is to be simulated.
2 Obtain probability distributions for each and decide the values that govern the shape and skewness of the probability distribution. (For example, for a triangular distribution three values – minimum, maximum and most likely – are sufficient.)
3 Allocate sets of numbers to various value ranges in the distribution.
4 For each simulation generate random numbers (repeat for desired number of simulations).
5 From the simulations, calculate the likely values of the variable at $x\%$ confidence.

The simulation part of the process is carried out in steps 3–5. Step 3, the allocation of sets of numbers may, for example, involve allocating 10,000 numbers (0 to 9,999) to the different possible values of the variable. If, from the probability distribution, there is, say, a 10% chance of a job being completed within 5 weeks, then an appropriate band of numbers (0 to 999) are allocated to that value. Once the numbers are allocated, simulation is carried out by means of a random-number generator. Producing random numbers is meant to simulate real-life chance, such as the results produced by a roulette wheel. (For this reason the process of simulating chance by generating random numbers is often called a *Monte Carlo simulation*.) The simulation is repeated for a desired number of rounds and the results are recorded. More repetitions will result in the simulated outcomes being closer to the original theoretical probability distribution.

6.3.3 Risk allocation and ownership

Having identified, analysed and evaluated the risks, it is a matter of managing them. Sometimes, however, when an organisation or individual is dealing with other parties, it may be possible to *allocate* risks to one of these other parties.

Some organisations/individuals are extremely risk-averse, and their immediate reaction is to dump all possible risks on others. There is nothing wrong with this provided that the party realises that normally it will be paying a price (a *risk premium*) for doing this. A commercially aware party will calculate whether it is better to keep the risk or 'sell' it. The usual method of risk transfer is by means of a contract. As already noted in Chapter 2, this is one of the construction contract's prime functions.

Risk allocation through contracts

As discussed in Chapter 2, the normal principles of efficient risk transfer suggest that a risk should be allocated to the party

- that is in the best position to bear the risk
- that wants to bear it (for some commercial or technical reason) or
- that has the greatest incentive to manage and mitigate the risk.

In some industries, buyers develop an intimate knowledge of sellers' characteristics, and this informs their decisions on risk allocation.

Example: Japanese motor manufacturers

Studies have revealed that Japanese motor manufacturers actually remove risks from their subcontractors rather than, as had been thought, pass the risks on.[8] The reason for this is that the subcontractors, being smaller, were more risk-averse than the larger 'generals', and had less capacity to bear the risks. By removing risks, the general contractors got a better price.

It could be argued that the drafters of traditional construction contract forms have tried to do this too. They purport to 'know' the industry, in the way that the Japanese general contractors know theirs. Thus the 'average' builder

- needs cash flow to survive – hence the system of interim payments to improve cash flow for the contractor
- is vulnerable to resource shortages – hence provisions such as JCT 25.4.10.1 and 2, which remove from the contractor the risk of liability for damages if he is (unforeseeably) unable to obtain labour or materials for the works.

Perhaps these standard forms are adequate for run-of-the-mill projects. But what about larger, more complex and more significant projects? More than 70% of the JCT contracts that are used get amended. So how should these projects have their risks allocated? Should it be done unilaterally, in isolation from the party whose capacity and incentives are crucial to the decision, or bilaterally, where the capacity, incentives, risk attitude and risk premiums can be openly taken into account?

In general, optimal risk assignment involves allocation of a particular risk to the party that can deal with it most efficiently. This is a particularly crucial issue in an industry such as construction, where risks are so prevalent because of construction's complex and uncertain environment. Here the form of contract has a particularly important role to play. There are a number of common clauses that are chiefly involved with risk allocation.[9]

Risk transfer through insurance

Of course, some organisations exist purely to 'buy' risks: these are insurance companies. Thus the second method of achieving risk transfer is through insurance. Insurance 'works' because of the combination of

- the fact that risk-taking comes at a cost – the risk premium
- the principle of 'large numbers' in statistics
- the pooling of risks.

Example

Consider a company that owns a car. Assume that it is possible to estimate the likelihood of that car suffering accidental damage. Say that on average there is a 1 in 1000 chance of a car having an accident costing £1,000 to put right. When the company owns only one car there is a degree of uncertainty surrounding the possibility of a loss. Typically a company owning one car would insure against the loss. If the company owned 1000 cars (or more) the likelihood of the 1 in 1000 chance of a £1,000 accident would increase. At a certain point the company would become a self-insurer to avoid paying a risk premium. This is precisely what the insurance company does by pooling the risks of a large number of small risk-holders, thus converting uncertainty into calculable risk. The insurance premium includes the cost of covering this risk plus a mark-up for profit.

6.3.4 Management of residual risk

When a party has evaluated the risks, and has decided which to allocate and which to retain, the retained, or *residual*, risks must then be managed. Managing risks involves continually monitoring and controlling the circumstances to try to ensure that the procedures that have been put in place to counter the risk are sufficient.

6.4 Conclusions

This chapter has looked at the techniques and processes of risk management. Construction risks have always been recognised, and allowed for, but the use of more formal, probabilistic techniques is relatively recent.

The following are the key points covered in this chapter:

- There is a common, generic process for risk management.
- It involves the stages of identification, analysis, evaluation, allocation and management.

- Techniques have been presented for identifying risks.
- Techniques have been presented for analysing and evaluating risk.
- The allocation of risk is normally performed through contracts or through insurance.
- Risks that are not transferred (that is, residual risks) must be managed: usually this involves controlling them.

Notes and references

1 By thinkers such as Pascal, Fermat, Bernoulli, De Moivre and Bayes.

2 Such as the economist Jevons.

3 Primarily Von Neumann and Morgenstern.

4 For a serious but enjoyable history of the development of risk theory see P.L. Bernstein *Against the Gods: The Remarkable Story of Risk* (John Wiley & Sons, New York, 1996).

5 The approach – combining the probability of a hazard with its impact – is often referred to as the Bayesian method, after the eighteenth-century English mathematician Thomas Bayes, though in fact the idea predates Bayes' work.

6 Named after the oracle at Delphi, in Ancient Greece. See (e.g.) R. Graves *The Greek Myths,* Vol I, pp 178–182 (Penguin Books, 1955; combined edn Carcanet, Manchester, 2001).

7 CIRIA *Control of Risk: A Guide to the Systematic Management of Risk from Construction* (Construction Industry Research and Information Association, London, 1999).

8 See S. Kawasaki and J. McMillan 'The design of contracts: evidence from Japanese subcontracting' *Journal of the Japanese and International Economies* Vol. 1 (1987), pp 327–349..

9 There are several pieces of related research. Two are from the United States: C.A. Erikson and M.J. O'Connor *Construction Contract Risk Assessment,* Technical Report P-101 (US Army Construction Engineering Research Laboratory, Champaign, IL, 1979); C.W. Ibbs, W.E. Back, J.J. Kim, D.E. Wall, J.M. DeLaGarza, S.M. Hassanein, S.M. Schran and R.K. Twardock RK *Determining the Impact of Various Construction Contract Types and Clauses on Project Performance* (University of Illinois at Urbana-Champaign, 1986).In the UK similar work has been undertaken. See: P. Hibberd, D. Merryfield and A. Taylor *Key Factors in Contractual Relationships* (Royal Institution of Chartered Surveyors, London, 1990).

7   Life cycle costing

7.1   Introduction

Life cycle costing is concerned with considering the total cost of a building over its total life. This goes beyond the initial construction costs and takes into account the costs arising from, for example, maintenance, repair, ownership and management of the building. In the case of whole life costs this includes the cost of recycling and demolition of the building – the cycle of costs that arise on the journey from green field to building and back to brown field.

Various factors have pushed the issue of total life costs of buildings up the agenda of the construction and property industries. There is, at the start of the twenty-first century, a greater concern with the well-being of the planet. This concern with *sustainability* has led to a more critical view both of the way buildings consume energy and of the energy involved in converting raw materials into building components and delivering them to site. Alongside this is a growing awareness of the impact of harvesting or extracting raw materials used in construction, and the impact of this on long-term sustainability.[1] This concern extends beyond the life of the building to the reuse, recycling and disposal of the components of construction when the building is demolished. Indeed, rather than demolition, increasingly consideration is being given to ways of dismantling buildings and recycling materials.

The need and the ambition to design sustainable buildings are increasing. Buildings consume about 50% of the energy generated in the UK. The construction and occupation of building contributes significantly to CO_2 emissions. The Kyoto Protocol, to which the UK government is committed, and the introduction of the Climate Change Levy are examples of changes in political, fiscal and regulatory policy in support of sustainability.[2] However, it is probably true to say that the take-up of sustainable measures in building design has been hindered by the failure to make a convincing commercial case for sustainable buildings. Life cycle costing is an important tool for demonstrating the real financial benefits of sustainable design.

In recent years a greater awareness has developed of the costs associated with owning and operating buildings, and in particular the balance between the cost of constructing a building – the *capital cost* – and the cost of owning or operating the building – the *cost in use*. Estimates suggest that the capital cost typically will account for only 40% of the whole life cost of a building; the cost in use accounts for the other 60%.[3]

In earlier chapters we have looked at the way different ways of purchasing and owning buildings can influence attitudes to risk, value and cost. A good example of this is buildings procured under the *Private Finance Initiative* (PFI). In a PFI project the consortium will not only build the building but will also be responsible for the ownership, management and maintenance of the completed facility. The consortium receive a 'rent', or some similar monthly income, for delivering the building, and hence the running costs incurred will be significant in terms of reducing the ultimate income stream – and therefore the profit. Hence there is likely to be a far greater emphasis on reducing maintenance, energy and management costs from the outset, by design.

A further example is in *partnering* arrangements, which bring the owners and the operators of buildings into the procurement process much earlier than was once the case. This also brings to the design process a knowledge and an awareness of operating and ownership costs and how these can be reduced by design.

The availability of software both to carry out a sophisticated analysis of (for example) the thermal performance of the building, and to run financial projections and sensitivity analysis, has served to make the examination of life cycle costs more accessible and potentially more accurate. However, inevitably the problem is not with handling and manipulating the data but rather with obtaining reliable input data, and with making judgements about forecasting factors and interpreting the output results.

Whole life costs include all the life cycle costs, and take into account the construction costs and the cost of extracting raw materials for the building components and transport costs of materials for construction. They also take into account the demolition costs and the cost (or income) from recycling or disposing of the building materials.

7.2   Costs incurred in using buildings

Typical costs arising from a building in use include the following:

- *Energy costs.* These include the cost of heating, lighting and powering the building.
- *Replacement and maintenance.* This is the cost of replacing damaged or obsolete components, together with the cost of maintaining the fabric, the fixtures, the services and the equipment.
- *'Soft' facilities management.* This is the cost of staff to manage, service and maintain the building. A significant cost in this category would be the cost of

cleaning the building. This would also include recurring financial costs such as insurance and rates.

- *Recycling costs.* To be a truly accurate measure, the whole life cost should also include the cost of recycling, recovery and the salvage of components for reuse.

7.3 Life cycle costing: the component parts

At its most basic, life cycle costing requires consideration of the life of the building and of the costs that will be incurred during the building life. Expressed as a simple equation:

life cycle costs = life of building × running costs/annum

It is of course not quite as simple as that. To establish the first part of the equation – the life of the building – we need to consider the factors that will determine the lifespan. These can be considered in two categories: *obsolescence* and *deterioration*. For the second part of the equation – the cost – we need to consider the effects of *inflation* and the changing value of money over time, for which *discount rates* are used.[4] It is also necessary both to quantify the costs and to convert them into a common and manageable form.

To add one final complication, it is also prudent to take into account the impact of future changes on the variables in the equation. For example, a change in interest rates could have a significant impact on cost and therefore influence design decisions. The effects of these changes are normally tested by applying a *sensitivity analysis* to the calculations.

There can be no certainty in forecasting the future. As Sam Goldwyn famously said:

Forecasting is a difficult business – particularly about the future.

Forecasting requires appropriate experience, judgement and knowledge together with access to reliable data. However, it is often not necessary to arrive at *absolute* answers; an evaluation of the *relative* benefits of one option over another is often enough to allow a considered choice to be made.

7.3.1 The physical life of the building: deterioration

Deterioration can be considered in two categories: the physical decay of the building fabric, and the physical decay of individual components. An immediate complication can be seen in that the many components that make up a building will decay at differing rates.

Data on component decay are available and are published in various forms: for example in the *HAPM Component Life Manual*,[5] which is widely used by housing associations. This groups components into elements – floors, walls, mechanical equipment and so on – and within these provides typical specifications for component subtypes. For example, in the Walls section there is a subsection for timber boarding as a wall cladding system. Specifications are given for different types of timber cladding material. Each of these specifications is allocated a code, which includes a letter designating the life of the construction: B = 35 years, C = 30 years, D = 25 years, and so on. For example, *permeable softwoods dip-treated with organic solvents* are noted to have an expected life of 20 years and will require redecoration every 4 years. The cost of redecoration and of replacement therefore needs to be allowed for on this basis. A worked life cycle costing example using these criteria is given later in this chapter.

Significantly the manual recognises that the location of the component within the building may affect its performance, and also assumes that components are installed in accordance with manufacturers' guidance and that the works conform with British Standards and with good construction practice. The manual specifies the required maintenance and – in defining the component life – assumes that this maintenance will be carried out as a minimum. Adjustment factors are included to allow for local conditions: for example a marine or polluted environment.

Various factors will affect the rate of deterioration of components:

- The *quality of the materials* originally specified. For example, in a corrosive environment stainless steel will last longer than galvanised steel, which in turn will outlast mild steel.
- The *design and detail* of the building. Inappropriate detailing can for instance lead to accelerated weathering, which will reduce the life of components. For example, a failure to detail the capping and damp-proofing of a brick parapet wall properly may lead to the bricks' becoming saturated, and in cold weather the freezing of entrained water may cause the brick face to spall.
- Poor *construction practice* in assembling the building, for example where materials are not properly protected prior to making the building weathertight or where equipment is damaged in installation.
- The life of components can also be affected during the use of the building by inappropriate use, or more correctly, *abuse*. Floors that are designed for foot traffic may have fork-lift trucks passing over them from time to time, with a consequent detrimental effect on the long-term life of the floor.

Example: the deteriorating cladding

An architect specified corrugated steel cladding for use on the external walls of a railway station building in a coastal location. The canopy over the platform protected the cladding from the effects of rain, although airborne salt and moisture-laden air could settle on the cladding. The design did not allow rainwater to wash the salt deposit from surface of the cladding, which meant that the owner had to hose down the cladding on a regular basis to avoid a build-up of corrosive salts. The stringent safety requirements for working alongside an active track make maintenance work costly and disruptive for a railway operator. In fact a more expensive cladding material requiring less maintenance might, when considered over the life of the building, have been less costly.

In some cases it may be appropriate to use components that last for a shorter time and hence avoid paying for redundant longevity. Building components deteriorate at quite different rates; there may be little point in installing long-lasting, more expensive components alongside materials that will deteriorate more quickly, as they will not increase the overall life of the building. It is only when a proper life cycle cost analysis is carried out that decisions such as this can be made in a considered way.

Ultimately, however, physical deterioration of the building is unlikely to occur before the building becomes obsolete.

7.3.2 The useful life of buildings: obsolescence

The physical fabric of a building that has been well designed, constructed and maintained may continue indefinitely. However, the use of the building – the function for which it is (or more precisely *was*) required – may disappear or reduce to the point where there are too many of a particular type of building. When the building can no longer fulfil its original function – or when the function no longer requires a building – then the building is obsolete.

In 1986 the RICS published *A Guide to Life Cycle Costing for Construction*,[6] which helpfully categorises the various ways in which buildings can become obsolete. These are as follows:

- *Technological.* This occurs when changes in technology mean that the building can no longer fulfil its function. This has been a common phenomenon in recent years with office buildings, where the considerable increase in the use of computers has led to a need for greater distribution of cabling. Modern office buildings are designed with access floors to accommodate this. The ability of older office buildings to be adapted is often dictated by the floor-to-floor height: if this is adequate to accommodate an access floor the life of the building can be extended. If not, the building will either become obsolete or will be let to a tenant on very poor terms.

- *Functional.* Changes in markets and social behaviour can lead to buildings becoming obsolete. This can be due to long-term structural changes in the economy – for example the decline of shipbuilding in the UK, making shipyards obsolete – or changes in consumer behaviour – the move to supermarket shopping and the demise of the corner shop. Even short-term changes in markets can have a dramatic effect: the sudden drop in the price of computer microchips led to a number of recently built microchip manufacturing facilities becoming obsolete almost overnight. To compound matters, these were often designed as highly specialised buildings that were difficult to convert to alternative use.

- *Economic.* When the value of the land exceeds the income generated by the building the building is economically obsolete: for example, a warehouse in an inner city location where the value of the land for redevelopment exceeds the income from the warehouse use. From the owner's point of view it makes more sense to sell the building and relocate the warehouse to an edge-of-town location.

- *Social.* Changes in social patterns and lifestyles, for example the move away from inner city living for families, made large areas of housing obsolete, and high-rise flats in particular. To illustrate how cyclic these matters are, a move back to inner city living by young professionals has led to vacant offices being converted into apartment buildings in many UK cities.

- *Legal.* Changes in legislation can impact on the use of buildings. For example, considerable investment may be required to comply with more stringent environmental health regulations for food production. The cost of these may not justify the investment in an existing building, and it may be more sensible to move to new premises.

- *Aesthetic.* Fashion in building design changes, like all fashions. Offices built in the 1950s and 1960s are difficult to let, even when adapted to accommodate new technology. This is in many cases simply a reflection of the unpopularity of the building's appearance.

Except for the small number of cases where a building is designed to fulfil a specific temporary function for a fixed time (the Millennium Dome, for example), it is generally impossible to predict when a building will become obsolete. However, the risk or likelihood of obsolescence can be reduced by design, for example by designing loose-fit, adaptable buildings that allow capacity for expansion in terms of space, layout and construction. This could also extend to designing for contraction or for subdivision. Clearly, the inclusion of such a provision may increase the initial capital cost. For example, designing the structure so that it can accommodate an additional storey in the future may increase the structural frame costs, but when the increased useful life of the building is taken into account this premium may well prove worth paying. This is a matter of value judgement, and it is this type of question that life cycle costing can help to answer.

7.4   Cost and prices

7.4.1   Inflation

Inflation is the measure of the way prices for goods and services move upwards over time. High levels of inflation are considered to be a bad thing, as inflation causes money to lose its purchasing power. Sustained high levels of inflation ultimately create instability and uncertainty in markets, and for this reason much of government economic policy is concerned with controlling inflation.

The most common measure of inflation in general use is the Retail Price Index (RPI), but specific construction measures – the Tender Price Index and the Building Cost Index – are more appropriate measures in the context of life cycle costing. They are available from a variety of sources: for example, the RICS Building Cost Information Service (BCIS) provides monthly updates, which are available by subscription.[7]

Weekly construction publications also include data that are updated on a regular basis. The information is presented both as historic data and as future projections. Projections are arrived at by producing economic models based on historic trends and a view of future trends. Generally the further into the future the projection is, the less reliable it will be.

7.4.2   Discount rate

In order to compare costs now with costs in the future it is necessary to apply a *discount rate*. This represents the opportunity cost of capital: that is, the best return that could be obtained on an alternative use of the funds. It can also be arrived at by reference to the cost of borrowing either from a firm's own funds or reserves (in which case interest that would have been received is lost), or from external sources (in which case interest is paid). The discount rate can therefore be thought of as compound interest in reverse.

In life cycle costing, discount rates are used to calculate the equivalent present value of a sum of money to be spent in the future, so that capital sums – money to be spent now – can be compared with projected maintenance and replacement costs. For example, £100 invested at 10% per annum compound would after 2 years grow to £121. On the basis of the ability to invest money at 10%, therefore, £121 in 2 years' time would be worth £100 today: so if you are committed to spending £121 on redecoration in 2 years' time you need only invest £100 now.

The formula for calculating the present value can be expressed as follows:

$$P_0 = \frac{Y_n}{1 + i} \; n$$

where P_0 is the present value; Y_n is the undiscounted value of cash flow at the end of the nth time period; i is the interest rate, expressed as a decimal; and n is the number of time periods between the present and the occurrence of Y_n.

The arithmetic may be straightforward, but the important point to note is the phrase *the ability to invest money at 10%*. For all sorts of reasons that ability may not exist: interest rates may go down, and therefore – as with all projections – a considered view has to be taken on the rate to be applied. There is no getting away from the judgement element of this, or indeed from the risk of making a decision based on assumptions that eventually turn out to be incorrect. However, the techniques of risk management, described elsewhere, are equally applicable in this area.

It is common practice often to ignore the effect of inflation in calculations, although strictly this will affect the result. This may sound inaccurate, but as what is often being sought is a comparative rather than an absolute answer, provided all other things are equal the 'answer' (that is, the decision taken) will be broadly similar whether inflation is taken into account or not. It will usually be accurate enough to assist in making a decision.

However, in some contexts this may be misleading. For example, to reduce energy costs the thermal performance of the building may be increased by providing a thicker, better insulated external wall. But this may reduce the net lettable area of the building and so reduce the rental income of the building over its life. In this case, for an accurate comparison it may be necessary to allow for potential differences in the rates of inflation of energy costs and of office rents. There is no reason why the two should move at the same rate.

The procedures for calculating life cycle costs are in many ways more difficult to describe than to do. A simple illustration using as an example the softwood timber wall cladding described earlier will demonstrate this.

Example: A simplified life cycle costing calculation
The *HAPM Manual* describes permeable softwood boarding as having a life of 20 years and requiring redecoration every 4 years. Using as an example a

traditional house, we can compare this with a traditional brick outer skin, which the manual defines as having a life of 35+ years, and which requires no maintenance apart from repointing after 20 years.

We shall assume (conservatively) that the house has a life of 40 years, and we shall assume an interest rate of 3% per annum compound. The resulting discount rates would typically be taken from published tables, such as those in *Parrys' Valuation and Investment Tables.*[4] Assume that timber cladding has an initial (capital) cost of £10,000 and that the cost of redecoration is £1,200. The alternative of cladding the house with brick costs £18,000, and the cost of repointing this after 20 years is £3,000.

Softwood timber wall cladding: life cycle costs

	£
Initial cost	10,000
Present value at 3% of:	
£1,200 redecoration cost after 4 years @ 0.88	1,056
£1,200 redecoration cost after 8 years @ 0.79	948
£1,200 redecoration cost after 12 years @ 0.70	840
£1,200 redecoration cost after 16 years @ 0.62	744
Renewal after 20 years (£14,000) @ 0.55	7,700
£1,200 redecoration cost after 24 years @ 0.49	588
£1,200 redecoration cost after 28 years @ 0.44	528
£1,200 redecoration cost after 32 years @ 0.39	468
£1,200 redecoration cost after 36 years @ 0.35	420
Total life cycle cost	23,292

Brick wall cladding: life cycle costs

	£
Initial cost	18, 000
Present value at 3% of:	
£3,000 repointing cost after 20 years @ 0.55	1,650
Total life cycle cost	19,650

So the traditional low-maintenance brick cladding is the less expensive option over the life of the building, even though it has a higher initial cost. There are of course other matters to take into account, such as personal taste in the appearance of the building, the disruption of redecoration, the ability to obtain planning permission for either cladding, and the likelihood of interest rates remaining at the low level of 3%. These are matters of judgement, to which the calculation can bring an objective and quantifiable dimension.

The calculation can be adjusted to take the effects of different criteria into account (see section 7.5 below). For example, if Western Red Cedar is used as the timber cladding this lasts longer (25 years), and requires redecoration only every 5 years, at a cost of £1,000. We can also examine the impact of an increase in interest rate to 4%.

Western Red Cedar wall cladding

	£
Initial cost	10,000
Present value at 4% of:	
£1,000 redecoration cost after 5 years @ 0.83	830
£1,000 redecoration cost after 10 years @ 0.68	680
£1,000 redecoration cost after 15 years @ 0.56	560
£1,000 redecoration cost after 20 years @ 0.45	450
Renewal after 25 years (£12,000) @ 0.38	4,500
£1,000 redecoration cost after 30 years @ 0.31	310
£1,000 redecoration cost after 35 years @ 0.25	250
Total life cycle cost	17,580

Even at a higher rate of interest the Western Red Cedar, which has the same initial cost as the softwood cladding, is a more cost-effective option than either this or the brick cladding.

7.5   Sensitivity analysis

The simple example above illustrates the ability to apply a sensitivity analysis to the calculation. Sensitivity analysis is a well-established management quantitative tool. It involves running the same calculation with different values and permutations of values. For example, as shown, interest rates can be varied, and the effect of this on the selection of alternative specifications can be considered. This is of course simply another form of risk management.

As with all forecasting it is simply a view that is being taken as to how interest rates and inflation will rise and fall in the future – there are no guarantees. The future cost of energy and the relative costs of alternative forms of energy are also unknowns. Governments may use fiscal policy to encourage or discourage the use of types of energy or the amounts of energy consumed. Changes in tax benefits or the availability of grants towards capital or running costs must also be considered. It is estimated that the cost of electricity from solar power at present is about five times that of electricity from the national grid. But who can say how this will move if economies of scale in, say, the production of photovoltaic panels take effect, or if the government introduces financial incentives to encourage the use of sustainable energy?

7.6 Cost in use

7.6.1 The design process

In the traditional design process the quantity surveyor provides a cost advisory service to the design team. Rather like the way a management accountant advises a company, the QS provides the analysis and data on which the design team can make informed decisions. In order to establish the life cycle costs it may be necessary to extend the circle of advisors to include the facilities manager and the significant supply chain companies. For example, the supplier of the lifts for a building is well placed both to maintain these and to provide initial advice on the maintenance options and their associated costs.

One of the main advantages of life cycle costing is that it helps and encourages designers to make better, sustainable buildings where design choices are made on the basis of an objective cost analysis. However, the techniques involved are difficult because of the timescales involved: the life of buildings requires *long-range forecasts*. A further problem is that the analysis of cost in use is a relatively new area. There is a limited body of knowledge, and limited accumulated data. It can take a long time to gather the data associated with the life cycle cost of a building.

Life cycle issues must be considered at the outset of the decision to build. Decisions regarding plan form, site layout and the building organisation will all have an impact on the life cycle costs, and should be considered when the strategic brief is being written. Even the very early decisions such as the selection of the site will have an impact: consider for example the maintenance costs or the design issues involved in a building on site in a harsh marine environment.

In practice, a cost in use analysis involves the traditional quantity surveying techniques of measuring and applying rates. However, the difficulties of gathering data, and the relatively longer timescales involved in the total life of a building compared with the design and construction stage, mean that the quantitative techniques involved are likely to yield relative rather than absolute answers. Nevertheless, these are useful in informing design decisions.

7.6.2 The energy issue

For most buildings the biggest cost incurred in the building's life is the energy cost – the cost of heating, cooling, lighting and powering the facility. Experience suggests that the energy consumption can account for up to 30% of the total life cycle cost of the building. These costs are influenced both by the use of the building and by the actions of the occupants. They are also significantly influenced by the initial design decision. Increasing the thermal performance of the building by increasing the amount of insulation (the U-value) will reduce the running costs but will also inevitably increase the capital costs. Reducing the number of windows will reduce the heat loss but may increase the need for artificial ventilation and lighting, and may affect the occupants' enjoyment of the building.

In all of this, assumptions must be made regarding the likely cost of energy in the future. Increasingly, issues relating to energy usage are influenced not just by cost but also by concerns with sustainability or with the occupants' ability to control their own comfort within a building. Striking the correct balance at the design stage is complicated by the number of interdependent variables, the need to predict future costs, the absence of robust (or sometimes of any) historic data, and the particular aspirations and philosophy of the building owner, developer or occupier.

To address these issues successfully the design team, and in particular the building services engineer and the architect, must work closely together at the outset to examine the implications of alternative designs. Computer software can be used to build a model of the building, and the effects of increasing the areas of glazing, orientating the building in a different direction, or increasing the natural ventilation can all be modelled and costed in both capital and running cost terms.

7.7   Life cycle costing: uses and limitations

Life cycle cost analysis can be used to:

- *Compare alternative building design strategies.* This can include comparing the effects of changing the form, the orientation or the planning of the building.
- *Compare alternative buildings.* Where a client is contemplating buying a building an analysis can be performed to compare the life cycle costs of different buildings. This can be done even before the buildings are built.
- *Compare alternative sites.* The site – for example the topography, proximity to the coast, relation to surrounding buildings – can have a significant impact on maintenance or energy costs.
- *Compare alternative energy strategies.* Life cycle costing can be used to test the benefits of passive or active energy design: for example the advantages of increasing the level of insulation of the building measured against savings in energy costs over the life of the building. The costs of alternative types of energy can also be compared.

In practice, the comparison of alternative design strategies to conserve fuel consumption is a complicated process, as the following example shows.

Example: A tricky question – to glaze or not to glaze?

A firm of engineers examined the implications of alternative design strategies for an office building in London's West End. Reducing the area of glazing increased the insulation of the building and so reduced the need for air conditioning to deal with solar heat gain. However, this increased the need for artificial lighting, reduced the ability to make use of natural ventilation, and was likely to be unpopular with tenants and would therefore reduce the marketability of the building. Planning permission was also likely to be a problem. Further consideration that took into account the shading effect of adjoining buildings and the use of solar reflective triple glazing made the glazed option appear more attractive. This had to be balanced against the much higher capital cost, the increased cleaning cost, and the loss of lettable floor space due to the increased zone required for the triple glazing and its support structure.

All these questions arose before consideration was given to the many imponderable forecasting issues. How would energy prices move in the future? What levels of rental would be achievable for London office space? How would this grow or decline? What impact would the demolition of the adjoining building have on the beneficial shading effect? How likely was this? How would the occupants use the building – would they close blinds or let the air conditioning work harder? Would the building be well maintained?

The challenge in life cycle costing is to consider all the issues using the best available data with the most realistic projections. There are no absolutely right answers – only time will tell!

Life cycle costing has a number of limitations:

- *Availability of data.* The amount of available data is limited, although this is increasing as the use of life cycle costing becomes more widespread. Various publications that provide data on materials in use are now available. The data are often limited in practice by the size of the sample. There are no empirical data available on new, untried materials or forms of construction.
- *Take-up and application* (see section 7.9). Clients are often not aware of the benefits of carrying out a life cycle cost analysis. The take-up tends to be restricted to large projects with enlightened clients who intend to occupy the building themselves (as opposed to developers building for sale). However, as the cost savings should more than cover the cost of a consultant preparing a study, there is no reason why it should not be done on most projects as part of the normal cost planning process.
- *Reliability of assumptions.* All projected costs are subject to uncertainty. Although the cost of alternative fuels for energy will tend to balance out, over time there will be movement in the relative cost of each. This may mean that

assumptions made at the outset do not hold true for the life of the building. A sensitivity analysis will allow the testing of alternatives to examine their relative ranking. As with all financial projections the discount rate chosen at the design stage may be affected by changes in actual interest rates and inflation. This can also be addressed by a sensitivity analysis.

7.8   Whole life costs

Looking beyond life cycle costing there is also increasing concern to measure the entrained energy within the materials used in constructing a building. This has extended to attempts to measure the reuse, demolition and disposal cost associated with alternative materials. This is referred to as *whole life costs.*

The Kyoto agreement has already given rise to changes in the Building Regulations (changes to Section L, 'Conservation of fuel and power'[8]), with the aim of reducing energy consumption. It is likely that such concerns will continue to grow. Cost in use analysis provides the rational and objective means to test the performance of buildings in this regard. It is important, however, not to consider these results in isolation. Viewed in the wider economic context, there is little point in reducing the energy consumption of a building if the savings made are exceeded by the energy consumed in making the insulants.

The biggest challenge will be to bridge the gap that seems to exist between theory and practice. Although there is an increased awareness of and concern with the profligate use of energy and materials, there is still a surprisingly low take-up of cost in use studies as a practical process on everyday projects. Everyone sees the benefit in theory, but few are willing to translate this benefit into a practical application.

7.9   The take-up of life cycle costing in practice

Life cycle costing is a potentially powerful project decision-making tool. Why then is its take-up so low? There are various reasons, as discussed below.[9]

7.9.1 Lack of client motivation

Clients are often not prepared to pay for a life cycle costing exercise. Three reasons are suggested for this:

- There is a lack of confidence in the accuracy and therefore the usefulness of the results.

- A limited understanding of life cycle costing has led to a general failure to appreciate its benefits. Property developers perceive the benefits as being for the occupier, not themselves: therefore why should they bother? This misses the point that a building with low running costs may let more quickly and on better terms than one with high running costs.
- The process is seen as artificial: no one actually puts away the sums of money to be spent for future expenditure, but the present capital expenditure is all too real.

7.9.2 Contextual issues

- Design teams will generally not volunteer to carry out a life cycle costing analysis unless the client commissions it and pays for it. It is an additional rather than a normal design team service.
- Factors other than running cost will play a far bigger role in deciding the design strategy.
- The split between the control of the 'capital' and the 'revenue' budgets in many public and private organisations can militate against taking a holistic, 'joined up' view of the two areas of cost.

7.9.3 Methodological limitations

- A relative study of alternative design strategies is a straightforward process. An in-depth quantitative study to give absolute answers is both time consuming and costly.
- There is a lack of common methods for life cycle costing and therefore it is difficult to share information and develop techniques.
- The bespoke nature of buildings makes it more difficult than in other industries to apply life cycle costing to analyse finished products. Compare for example the widespread availability of data on and common methods for measuring car fuel consumption with the lack of any equivalent data for buildings.

7.9.4 Access to reliable data

- Operational cost data are difficult to obtain.
- Life cycle performance information is limited, and not available for many materials and forms of construction. Sustainable construction often involves the use of innovative materials or forms of construction – precisely those for which data are not available.

Despite all these problems there are real benefits to be obtained from life cycle costing, which need to be communicated in order to encourage a greater take-up of life cycle costing in practice. It will also be necessary for the construction industry to invest in the accumulation and dissemination of life cycle cost data.

7.10 Conclusions

Life cycle costing seeks to establish the total cost of a building across its entire life. This requires consideration of:

- the useful or functional life of the building – the issue of obsolescence
- the physical life of the building – the issue of deterioration
- the projected costs arising from energy consumption, replacement of components, maintenance, facilities management, recurring financial costs (rates, insurance, etc.) and, for whole life costs, demolition, disposal and recycling.

These projected costs have to be adjusted to take into account:

- the effects of inflation
- changes in the value of money over time.

This is done by applying a discount rate.

A growing awareness of the relationship of life cycle costs and capital costs (and in particular the realisation that the former will probably exceed the latter), together with a greater concern on the part of individuals and government with the long-term sustainability of the planet, are likely to lead to an increasing use of life cycle costing as a tool in the building design process.

Design decisions made at the very outset of a construction project can have a significant impact on the life cycle costs of a building.

Problems of low take-up and use in practice include:

- lack of motivation, related to a lack of confidence in the results
- contextual issues, bigger issues than life cycle costing determining the design strategy
- methodological limitations
- difficulty of access to reliable data.

As with value management and risk management the problem is often to persuade clients to carry out a life cycle costing exercise. This can often be overcome by demonstrating the real financial benefits that arise. Reference to case studies of completed buildings is often the most persuasive tool.

Notes and references

1 There are literally hundreds of books dealing with the subject of buildings and sustainability. A good reader that covers a wide range of topics on this subject is I. Abley and J. Heartfield *Sustaining Architecture in the Anti-Machine Age* (Academy Editions, London, 2001).

2 For further information on the UK government and sustainability issues visit the Office of the Deputy Prime Minister web site at www.odpm.gov.uk.

3 For further reading on the wider context of life cycle costs see R.R. Morton and D. Jaggar *Design and the Economics of Building* (E & F N Spon, London, 1995).

4 For discount rates (and much more) see A.W. Davidson *Parry's Valuation and Investment Tables,* 12th edn (Estates Gazette, London, 2002).

5 Construction Audit Ltd *HAPM Component Life Manual* (E & F N Spon, London, 1992).

6 RICS *A Guide to Life Cycle Costing for Construction* (RICS Publications, London, 1986).

7 BCIS *Surveys of Tender Prices* (RICS Publications, monthly).

8 DTLR *The Building Regulations Approved Documents* (The Stationery Office, 2001).

9 For further reading see *Building Research and Information,* special edition ed. R.J. Cole, *Cost and Value in Building Green,* Vol 28 (2000), Nos 5, 6.

8 The human and social aspects of risk and value

8.1 Introduction

So far, risk and value have been dealt with in isolation from the propensities of the decision-makers themselves. But even the most cursory examination of real human behaviour reveals that people's attitudes to risk and value can vary enormously.

When any individual makes a decision about risk or value there are two components:

- an *objective* component, in terms of the inherent risk or value of a particular thing
- a *subjective* component, in terms of the individual's evaluation of that risk or value.

Furthermore construction is not merely a technical process. It relies upon numerous social systems that comprise quite different participants with different attitudes. Negotiation, compromise and sometimes conflict are regular features of the way in which these social systems react and interreact. This introduces a third component to the basic decision-making model:

- the influence upon risk or value decisions of the *group or groups* that act as stakeholders in the decision process.

The first part of this chapter looks at the subjective component of an individual's decision-making and, in particular, at attitudes to risk and value. This is followed by a discussion of the behaviour of groups, and their influence on decision-making.

8.2 Expected value and the individual decision-maker

The concept of *expected value* (EV), which was introduced in Chapter 6, provides a very convenient way to assess alternatives. The EV is produced by comparing the product of the *outcome value* of each alternative and the *likelihood* of its realisation. Thus, for whatever the event, *risk* in the form of EV is given by

EV = probability of occurrence × magnitude of consequences

Although the conspicuous word here is 'value', precisely the same calculation – *impact × probability* – underlies virtually all risk assessment techniques as well.

Example

A 1 in 5 chance of realising £10,000 (EV = 1/5 × £10,000 = £2,000) is a
better option than a 1 in 10 chance of realising £15,000 (EV = 1/10 × £15,000
= £1,500) because in terms of probability, £2,000 is better than £1,500.

A problem with EV calculations

Unfortunately experience shows that in reality people do not always behave as the
theory predicts. Is there something lacking in the theory, or are humans, as *Star
Trek's* Mr Spock would argue, just incapable of rational behaviour?

Consider the following example. It is based on the same simple EV calculation that
we have seen before.

Example

A businessman has to decide whether to risk investing in a business venture.
The data at his disposal suggest that there is an 80% probability of a £20,000
gain, though there is a smaller (20%) possibility of a £50,000 loss. What should
he do? A quick EV calculation suggests that the investment should be taken. But
observations of analogous situations suggest that in many cases businesspeople
will *not* risk decisions that will permit even a small chance of a large loss.

The position taken up by these cautious businesspeople is referred to as *risk
aversion*. Risk aversion is itself one of a number of bands of *risk attitude*.

8.3 Risk attitude

Experimental work has identified that individuals, types and groups can exhibit
attitudes that can be classified as *risk-averse, risk-neutral* or *risk-taking*. The
circumstances that cause these attitudes are discussed later, but the following
simple experiment will illustrate their difference.

Example

Members of a game-show audience are invited to take part in a one-to-one
gamble with the show's presenter. The gamble is based on that simplest of
games – spinning a coin. The rules are that members of the audience have to
bid for the privilege of being the one who competes: the highest bidder enters
the contest with the presenter, and even gets to call the result – heads or tails
– of spinning a £1 coin. If the call is correct the bidder wins the coin. In either
case the bidder pays the amount he or she bid. The bidding starts at 10p and
rises in 10p increments. Bidders start to drop out along the way: at 50p there
is a significant fall-off in interest, leaving only a few bidders to continue
competing beyond that point. In extreme cases the bidding will go beyond £1.

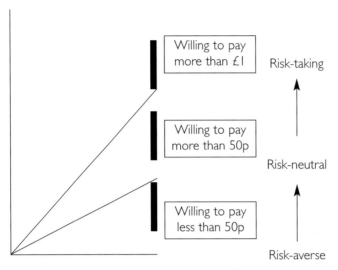

Figure 8.1 Three bands of risk attitude

What does this tell us? Let us start with the *risk-neutral* position. EV theory tells us that, since the *value* of the outcome is £1, and the *probability* of its occurrence is 50:50, then the EV of the gamble is 50p. The risk-neutral bid should therefore be 50p or just under. Those who dropped out of the bidding before the 50p threshold are exhibiting different degrees of *risk aversion*, while those who exceeded a 50p bid are showing increasing *risk-taking* tendencies. Figure 8.1 illustrates these results.

This example is both simplistic and unscientific: with sums as small as £1 we can be frivolous, and no doubt raising the stakes would produce different results. Nevertheless it illustrates the three positions fairly well.

The £1 bidding game and the cautious businessman in the earlier example demonstrate two points of great importance to professionals who deal in risk and value:

- The use of money as a proxy for value can be misleading. In the case of a cautious businessman the risk of a £1,000 loss outweighs the chance of a £1,000 gain. In other words two identical units of currency (£s) may have the same *face value* but very different *actual values*, depending on the circumstances.
- Risk and value calculations are affected not only by the circumstances of the situation, but also by the inclinations of the participants, who may be naturally (or calculatedly) averse from or inclined towards risk.

Unfortunately this is not the end of the complications that surround attempts to calculate value and risk, as the next section shows.

8.4 Utility theory and some further unexpected behaviour

The problems raised by the cautious businessman and the £1 bidders have been addressed by what is referred to as *utility theory*. This attempts to reassign a quantitative predictability to risk responses, and thus counter the problem of different risk attitudes. Instead of simply using expected monetary value in scales that measure preference, utility theory takes a more subtle approach to account for the willingness to take or avoid risk: *utility* supersedes monetary value, and the unit of measurement is the *utile*.

Figure 8.2 relates utility theory to the cautious businessman. Had he been risk-neutral there would have been no difference (apart from the obvious preference for gaining money rather than losing it) between the way he compared a £1,000 gain and a £1,000 loss. If, using Figure 8.2, we were to say that £50,000 was equal to 50 utiles, then in the case of the risk-neutral decision-maker we could say that £25,000 would correspond to 25 utiles, £75,000 would be 75 utiles, and so on. Note that this is not the case when the decision-maker is either risk-averse or risk-seeking. In the former case, small business gambles would appeal to our decision-maker more than large ones, while the risk-seeking businessman would be proportionally more attracted by the prospect of bigger rewards.

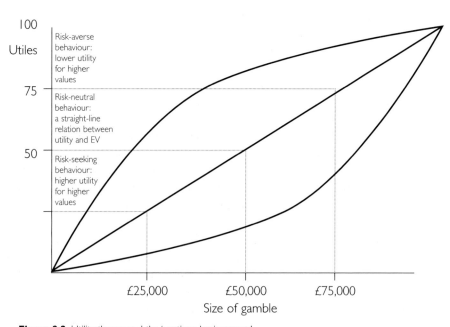

Figure 8.2 Utility theory and the 'cautious businessman'

8.4.1 The Allais paradox

Despite the contribution of utility theory, it is still deficient in practice. For example, the demand for mass public lottery tickets has been shown to be related not to the *expected* value nor even the *expected utility* of the prize, but simply to its sheer *size*, however great the odds against winning it. This type of decisional inconsistency is often referred to as the Allais paradox (after Maurice Allais, the distinguished French economist who first described it).

The following illustration again uses the example of a lottery.[1]

> **Example**
> Consider the following two sets of problems.
>
> *Problem 1: Do you prefer situation A or situation B?*
>
> Situation A: Receive a certain £1 million.
> Situation B: Receive a lottery ticket with:
> (a) a 10% chance of winning £5 million
> (b) an 89% chance of winning £1 million
> (c) a 1% chance of winning nothing.
>
> *Problem 2: Do you prefer situation C or situation D?*
>
> Situation C: Receive a lottery ticket with
> (a) a 10% chance of winning £5 million
> (b) a 90% chance of winning nothing.
> Situation D: Receive a lottery ticket with
> (a) an 11% chance of winning £1 million
> (b) an 89% chance of winning nothing.
>
> In Problem 1 the expected value of B is higher than that of A since $EV(B) > EV(A)$,[2] so rational people *should* prefer B to A. However, numerous empirical studies have shown that most people tend to prefer A to B, which goes against the EV logic. In Problem 2, the expected value of C is higher than D,[3] and indeed people tend to prefer C to D in practice.

Why do people go with EV logic in Problem 2, but not in Problem 1? The suggested reason is that people are lazy. When they can avoid a calculation, and instead use a *heuristic* (or rule of thumb), they will; they will indulge in calculation only when the heuristic no longer suits. There are thresholds at which this will apply. In the example, people choose A because they are using a heuristic that contains a strong preference for the *certainty* of A. In Problem 2 there are *two* uncertain situations with similar likely outcomes: the decision-makers can no longer rely upon the certainty heuristic, and have no choice but to 'cross the threshold' and make the calculation. Having got this far, they end up making the 'correct' decision (that is, using EV logic).

8.4.2 Framing and context of the decision

Research has also shown that not just the nature of the problem but also the way in which it is presented can have a significant effect upon decisions. Controlling the way a problem is presented is referred to as *framing*. The frame that a decision-maker adopts is controlled partly by the external formulation of the problem (that is, the way it is put) and partly by its subjective formulation by the decision-maker (that is, the way it is considered).

The following problem[4] illustrates the framing effect; we shall persist with the image of the lottery.

Example
Respondents were asked to make the following pair of concurrent decisions; to examine both decisions, and to decide upon their preferred option in each case.

Decision I
Choose between:
A   a sure win of £240
B   a 25% chance of winning £1,000, and a 75% chance of winning nothing.

Decision II
Choose between:
C   a sure loss of £740
D   a 75% chance of losing £1,000, and a 25% chance of losing nothing.

When confronted with these choices, 84% of subjects asked to make Decision I chose A. Only 16% chose B. With Decision II, however, 87% chose D and 13% chose C.

In terms of EV logic Decision I and Decision II are identical, except that Decision I concerns a gain and Decision II concerns a loss. In Decision I the choice is between a *certain gain* of £240 and an *expected gain* of £250.[5] Most people, quite naturally, chose A. With Decision II, C would bring a *certain loss* of £740, and D an *expected loss* of £750.[6] In this case, the majority chose D.

In this example of the framing effect, the majority chose a sure win of £240 (A) in Decision I because of a marked tendency for people to be *risk-averse* concerning *gains* (positively framed situations). By contrast, the majority chose a 75% chance of losing £1,000 (D) in Decision II because people tend to be *risk-seeking* concerning *losses* (negatively framed situations).

8.4.3　The 'availability' phenomenon

People are more deeply influenced than they realise by what they have recently heard on, for example, the media.

Example
When asked 'Which of the following causes more deaths in the USA each year: (a) stomach cancer or (b) motor car accidents?', most subjects believed that motor car accidents caused more deaths. This may not seem surprising, except for two facts about the year during which the survey was undertaken:

- Stomach cancer actually caused twice as many deaths as motor vehicle accidents.
- During the previous one-year period, two popular newspapers reported 137 stories that involved deaths by motor vehicle accidents and only one concerning stomach cancer.

It was argued that this went some way to demonstrating that the media coverage of the two phenomena had biased people's perception of the frequency of events, and thus their risk perception.

8.4.4　Statistical misconceptions

One of the causes of individuals' apparently 'irrational' decision-making is rooted in misconceptions about statistics. There are at least three ways in which these misconceptions can arise. The first has been referred to as insensitivity to *sample size*.

Example
A survey was conducted using the following scenario. A certain town is served by two hospitals. In the large hospital about 45 babies are born each day, and in the smaller hospital about 15 babies are born each day. About 50% of all babies are boys, but the exact percentage varies from day to day: sometimes it may be higher than 50%, sometimes lower. For a period of one year, each

hospital recorded the days in which more than 60% of the babies born were boys. Which hospital do you think recorded more such days?

(a) The large hospital
(b) The small hospital
(c) About the same?

In the survey, most people expected the two hospitals to record a similar number of days on which 60% or more of the babies born were boys. However, it is much more likely that 60% male babies will occur in a smaller sample than in a larger sample (just as you are more likely to get 6 heads from 10 spins of a coin than to get 600 heads in 1,000).

The second misconception is about *chance, sequences and randomness*.

Example
An estimator is putting in his fifth bid this year. He imagines that this one should stand a chance of success since the last five have failed, and he knows that his normal success rate is 1:5. Is his thinking correct or incorrect?

The answer (incorrect) also reveals why gamblers lose more money than they rationally should. People expect a sequence of random events to look random. For example, in a sequence of coin spins people imagine that

(1) H–T–H–T–H–H

is more likely than

(2) H–H–H–T–T–T

This is because (1) *looks* random, whereas (2) does not. However, each of these sequences is equally likely because of the *independence* of multiple random events.

The final example of misconceptions about statistics shows that people overestimate the true likelihood of multiple (*conjunctive*) events occurring, and underestimate the true likelihood of isolated (*disjunctive*) events occurring.

Example
Consider the following question. Which of the following seems most likely? Which seems the second most likely?

A   Drawing a red marble from a bag containing 50% red marbles and 50% white marbles.

B   Drawing a red marble seven times in succession, with replacement (a selected marble is put back into the bag before the next marble is selected), from a bag containing 90% red marbles and 10% white marbles.

C   Drawing at least one red marble in seven tries, with replacement, from a bag containing 10% red marbles and 90% white marbles.

The most common answer in ordering the preferences is B–A–C. Interestingly, the correct order of likelihood is C (52%), A (50%), and B (48%) – the exact opposite of the most common intuitive pattern! When multiple events all need to occur (B), people overestimate the true likelihood, while if only one of many events needs to occur (C), then people underestimate the true likelihood.

8.5   The influence of group behaviour

The previous section discussed the difference in risk attitudes of individuals. Individuals show differences in their conceptions of risk and value. But there are also differences between groups. These differences appear on two levels. First, all individuals are regularly classified into groups – by gender, age, race, etc. – and it is evident that these classificatory groups have differing risk and value attitudes. The fact that (a) women, are, in general, more risk-averse than men and that (b) there is an age factor to risk aversion become starkly clear when a parent first acquires motor insurance for a teenage son.

The second (and here more important) aspect of groups is that people join them and operate in them. The reason for this may be primarily social or economic: in the case of working life it is likely to be a mixture of the two.

8.5.1   Group cultural values and influences

Social groups have cultural values, to which members initially subscribe and over a longer timescale succumb. This is not only the case with groups that have formed for social purposes: the same is evident with groups whose primary function is economic. There is clear evidence that different professionals, working in different industries or sectors, will have different *mindsets*. This will include different concepts of risk and value. For example, a study of managers in banking, the oil industry and software concluded that there were clear industry mindsets when it came to risk strategies.[7] All three groups were confronted with the question 'What do you mean when you describe a business situation as "risky"?' The bankers' responses showed that they were, predictably, risk-averse, and that their main preoccupation was the level of control they could exert using known standards or criteria. Oil industry managers had a very calculative attitude: to them *risk* meant the extent of uncertainty of the whole range of outcomes. Software development managers, as well as being more open to taking risks, viewed risk not from the point of view of a particular outcome, but from the competitive position of their company – in other words, the risk of *not taking a risk*.

8.5.2 Groups acting together: the risky shift phenomenon

'Groupthink' and social mindsets are not just static phenomena; they can work dynamically whenever groups make decisions. When a group does this it will ultimately be involved in some kind of compromise – but the nature of the compromise can be surprising. Social psychologists in the early 1960s[8] first noted that groups are apparently more inclined to take risky decisions than individuals. Further results continued to support what had come to be known as the *risky shift phenomenon*: groups take riskier decisions.

There have been two basic explanations for this. First, individuals are willing to take greater risks collectively because they are removed from individual responsibility and therefore blame (the *diffusion* theory). Second, risk-taking is, up to a point, socially desirable, and risk-averse group members tend to emulate the risk-takers more than vice versa (the *risk is a value* theory).

8.5.3 Inefficient outcomes in compromises and negotiated decisions

When groups make decisions about risk and value, their decision-making becomes complicated by a number of factors. It is not merely the influence of *group culture* or the peculiarities of *group risk attitudes* that make group decisions a difficult area. A further phenomenon can be illustrated by the well-known scenario called the *Prisoners' Dilemma*.[9]

Example: the Prisoners' Dilemma
In the original version, two vagrants (Pete and Dave) decide to commit a crime but are caught nearby, and arrested. The police suspect them, but there is no evidence. They are kept and questioned in separate cells. Each one is told that if they both confess they will be charged with the crime, but that the police will recommend leniency. But each is also told: 'If your accomplice confesses, and you don't, he will get off lightly and you will receive a very severe punishment.' They also suspect that if they both keep quiet they will be charged with vagrancy anyway and receive a light punishment. What should Pete and Dave do? What *do* they do?

A first, rational, view would be that they should both keep silent. But would they? A game-theoretic analysis of this scenario suggests that both prisoners will confess. In the language of game theory, this prisoners' dilemma illustrates a two-sided game with high potential for a *non-efficient equilibrium outcome*. In other words, the *equilibrium* of the game tends to end up with both prisoners confessing, even though they could both be better off (that is, more *efficient*) by remaining silent. The reasoning is as follows.

Look at the situation from Pete's viewpoint. If, for whatever reason, he believes that Dave is going to confess, what would be his best strategy? The choice is between the very severe punishment (if Pete doesn't confess) and the moderately severe punishment (if he *does* confess). His best strategy therefore is to confess. On the other hand, suppose that Pete believes that Dave is not going to confess: what would be his best strategy? The choice is between the vagrancy charge (if Pete doesn't confess) and getting off lightly (if he *does* confess). Again, his best strategy is to confess. If we reverse the roles, the outcome is the same.

This is why the equilibrium of the scenario has both prisoners confessing, however irrational this seems. They would then both end up with a moderately severe punishment as opposed to a light one if they both kept quiet. The ability to cooperate would make each prisoner better off, but they can't! Thus in game theory parlance, the *equilibrium outcome* of the game is not *efficient*.[1]

8.6 Conclusions

Decisions about risk and value, like all decisions, depend not only on the external context (the objective facts about the decision in question), but also on factors that are inherent in the decision-maker. Not only do individuals differ in the way they make decisions, but there are also a number of phenomena that distinguish the decisions made by groups.

The following are the key points covered in this chapter:

- Risk- and value-based decisions can be highly subjective.
- Individuals do not always evaluate decisions using the logic of EV.
- The concept of EV does not always work with individual decisions in uncertainty.
- There are various phenomena that account for this, including people's framing, availability, and statistical misconceptions.
- Group behaviour is also influential in decision-making, from the point of view of group cultural values and influences.

- When groups act together there is a change in their risk attitude: the risky shift phenomenon.
- When groups or individuals make decisions there is a distinct possibility that, because of the circumstances they are in, they will fail to take the decision that gives the best outcome.

Notes and further reading

1 Many of the examples are based on the work of two leading theorists, Daniel Kahneman and Amos Tversky.

2 Calculations: EV of A, $E(A) = £1$ million x 100% = £1 million; $E(B) = (£5$ million x 10%) + (£1 million x 89%) + £0 million x 1% = £1.39 million.

3 Calculations: $E(C) = (£5$ million x 10%) + (£0 million x 90%) = £0.5 million; $E(D) = (£1$ million x 11%) + (£0 million x 89%) = £0.11 million.

4 Based on an experiment run by Tversky and Kahneman.

5 Calculations: $E(A) = £240$ x 100% = £240; $E(B) = (£1,000$ x 25%) + (£0 x 75%) = £250.

6 Calculations: $E(C) = -£740$ x 100% = $-£740$; $E(D) = (£1,000$ x 75%) + (£0 x 25%) = $-£750$.

7 A.L. Pablo 'Managerial risk interpretations: does industry make a difference?' *Journal of Managerial Psychology*, Vol. 14 (1999), No. (2), pp 92–107.

8 J.A.F. Stoner 'A comparison of individual and group decisions involving risk' Unpublished Masters thesis (Massachusetts Institute of Technology, School of Industrial Management, 1961).

9 The game, probably the most influential in the game theory literature, was devised by Dresher and Flood at the Rand Corporation and reported by Tucker in 1950: A.L. Tucker (1950) On jargon: the Prisoner's Dilemma, *UMAP Journal*, Vol. 1, p 101.)

10 An *equilibrium* outcome is the normal outcome of the game. An *efficient* outcome is one where there is no alternative outcome that the players would unanimously prefer (in this example, both getting a lighter punishment).

9 Risk and value in partnering and PFI

9.1 Introduction

For many years the procurement of buildings involved a well-defined linear progression from brief to design to construction. The players involved appeared on stage in the same order for every project: the client, followed by the design team, followed by the building contractor. In terms of status and power the hierarchy was much the same. However, recently various factors have changed this established order.

The construction industry has been going through a period of accelerated change. Much of that change can be attributed to the normal march of progress or *supply-side push* – the commercial and economic pursuit of greater efficiency. In individual organisations long periods of low profit, the high commercial risks inherent in contracting, and a failure to successfully add either shareholder or stakeholder value have forced both consultants and contractors to examine their management and delivery systems.

Alongside this has been the pressure of *demand-side pull* from customers to improve the construction industry. The Egan Report[1] and the Latham Report[2] – the former initiated by customers and the latter by government – both set out an agenda for change. Both reports called for a move away from the construction industry's characteristic one-off temporary project alliances and towards longer-term *partnering* relationships. These partnering arrangements should include the key suppliers and subcontractors, in order to harness the complete construction expertise of the supply chain.[3]

The *Private Finance Initiative* was established to encourage the private sector to put forward proposals for meeting public and social needs on the basis of output specifications. This allowed a move away from public sector direct capital expenditure on commissioning and owning property to an arm's length outsourcing and leasing arrangement for its building stock.

Both the supply side (the construction industry) and the demand side (building promoters) have taken up and developed these initiatives. The PFI, as a means of procuring public buildings, and partnering alliances both in the private and increasingly the public sector have grown steadily in recent years. This can be seen, for example, in the increase in local government spending (and allocated spending) that goes to PFI projects (see Table 9.1).[4] As demonstrated by the figures in Table 9.2, the trend is set to continue.[5]

Table 9.1 Local government spending and allocation (by funding Department)

Funding Department	1997/98 (actual) (£m)	1998/99 (allocation) (£m)	1999/00 (allocation) (£m)
Environment, Transport and Regions	149	200	250
Education and Employment	22	130	350
Home Office	41	80	100
Health	37	30	30
Law Courts	0	60	70
Total available	250	500	800

Table 9.2 PFI credits (2001–2004) by Department

Funding Departments	2001–2002 (£m)	2002-2003 (£m)	2003-2004 (£m)	Total (£m)
Education and Employment	450	550	650	1650
Environment, Transport and Regions	332	565	685	1582
Home Office	100	125	125	350
Lord Chancellor's Department	70	70	70	210
Health	40	40	40	120
Culture, media and sport	30	30	30	90
Cross-agency schemes	30	—	—	30
Total	1052	1380	1600	4032

Partnering and the PFI create a whole new set of project and design management issues, which have implications for both risk management and value management. To understand these it is necessary first to examine the processes involved in each.

9.2 Partnering theory and practice: a brief introduction

The construction industry has two key problems:

- the fragmented nature of the team that carries out the project, which exists at many levels – between client and contractor, between designer and builder, and between different consultant disciplines (often even when these are part of the same commercial organisation)
- the complexity of construction contractual arrangements, and the resulting adversarial relationships within these teams.

Partnering addresses both these issues by creating a team working framework in which mutual objectives are agreed and pursued, disputes are resolved at source, and risk and rewards are shared equitably. Linked to this is the pursuit of continuous and measurable improvement, with the lessons learned from individual projects shared between all the partners.

Central to any partnering arrangement is the *partnering agreement* or *partnering charter*. This covers the partners' objectives in terms of project delivery, team working and relationships, payment, quality standards and safety. The partnering charter is not a contract, and hence tended to be used alongside a standard contract. One school of thought held that it was invidious to have a contract for a partnering group, as this would acknowledge the absence of trust and encourage a return to the traditional adversarial relationships. However, the lack of practicality and the risk involved have caused a move away from this view, and the need for contractual formality is now widely recognised. In 2000 the Association of Consulting Architects published a *Partnering Contract* (PPC2000). This is a very unusual contract: not only does it combine partnering and legal contract principles, it is also designed for multi party use.

In 1998 the Reading Construction Forum, based at Reading University, published *The Seven Pillars of Partnering*.[6] This identifies the essential seven components or 'pillars' that support any partnering arrangement:

1   *strategy* – developing the objectives with the client
2   *membership* – identifying and assembling all the firms required, including the full supply chain (constructors, designers and suppliers)
3   *equity* – rewards for all based on fair prices and fair profits

4   *integration* – improving the way firms work together, and in particular increasing trust
5   *benchmarks* – setting measurable targets as a basis for continuous improvement
6   *project process* – best-practice procedures and standards
7   *feedback* – capturing and sharing learning as a basis for future improvement and development.

The report also identifies three generations of partnering:

- *First-generation partnering* or *project partnering* – partnering arrangements set up for a single project.
- *Second-generation partnering* or *strategic partnering* – partnering alliances set up between client, contractors, designers and suppliers to carry out a series of projects. Many of the major UK retailers, for example, have such arrangements in place to carry out their store development programmes.
- *Third-generation partnering*, although not widely used at present, envisages a construction industry that works closely with its customers to develop new, comprehensive packages of products. 'Products' in this context means not just the physical building; it extends beyond the bricks and mortar to include sourcing and providing land, the supply of plant and machinery, and the provision of facilities management and financial packages.

9.3   Partnering and value management

So if partnering answers many of the same problems that value management seeks to address, does value management have anything to offer in a partnering alliance? And if it does, how might the value management processes differ from those employed on a traditional project?

Value management examines the design team's assumptions, and questions whether the design best answers the client's objectives and meets their needs in a way that adds the most value. A key aspect of partnering (the first pillar, strategy) is the need to develop a comprehensive and intimate understanding of the client's business strategy and objectives. If this level of understanding is achieved, then there should be less need for the reappraisal that value management brings.

Value management emphasises the need for well-integrated multidiscipline involvement. As this integration rarely exists in traditional design teams, value management can lead to real improvements in the design. An essential feature of partnering is membership (pillar 2) of all the firms that need to be involved, with the people bringing all the necessary skills to the project. The more integrated teams are, and the more familiar they are with working together, the less the need for value management to bring about this multidiscipline approach.

Central to the value management process is the value management workshop. This integration of the team and the client can overcome the usually fragmented nature of the construction design process. In a partnering alliance there is a constant drive to encourage and bring about greater integration (pillar 4). Partnering alliances, because of their longer life, can afford to invest both in such things as project-based web sites (extranets) for the exchange of information between project team members, and in the time needed for team building.

Clients, consultants and contractors who are sufficiently committed to set up partnering arrangements are likely also to be committed to getting things right, and therefore arguably have less need for the benefits that value management can bring.

However, partnering is also concerned to use the best project processes. Even in the most efficient partnering alliance value management as a project process can add value by:

- creating the opportunity to briefly 'stop' the design and re-examine the project objectives in a formal workshop environment
- allowing people from outside the partnering alliance to bring a fresh set of eyes to the project and question both the strategy and the detail design
- focusing exclusively on the delivery of value.

9.3.1 Advantages of value management with partnering

Value management will inevitably differ in a partnering situation.

- Strategic partnering allows the project team to develop a long-term relationship with the building promoter, client or user. This can lead to a far greater understanding of the client's business needs, and hence a better understanding of the aspects that add value in the design and building.
- As in manufacturing, the same (or similar) buildings may be rolled out several times. Partnering works particularly well in this situation. Many retailers have standard stores that they repeat across the UK with only minor variations. Probably the best-known example is the McDonald's fast food chain. In such cases the cost of carrying out a detailed functional analysis can be justified; the return on investment comes when the buildings are repeated.

- The involvement of a contractor in the early stages of brief development and design allows the contractor's technical and management knowledge to be captured.
- The involvement of the supply chain in the early stages – the specialist suppliers and subcontractors – means that the value management workshop can include a better cross-section of participants.
- There is an opportunity for shared learning, both across partnering teams working on different projects and over time on a number of projects.
- It is possible to access feedback from the building user and from those involved in managing the facility.
- The opportunity to keep the project team together and to carry forward this expertise and experience of working together can make an important contribution to the maximisation of long-term value.

9.3.2 Problems with value management with partnering

There are also inherent problems in partnering. These include the following:

- If transactions are conducted without any element of competition, as for instance in a rolling strategic partnership where new projects are negotiated rather than tendered, there is always the risk of some form of opportunistic behaviour. In the absence of any form of competitive tender, it is impossible to be certain that the best price – which may or may not be the lowest price – has been obtained. Partnering usually works on an open-book basis, but it is often difficult to completely scrutinise the build-up to a price. There may be a strong element of trust (and a commercial interest) in not abusing the partnering arrangement, but it is the lack of competition that is likely to inflate prices rather than abuse of trust. The effect is of course the same: the price rises. High value does not necessarily equate to lowest price, but anything that artificially inflates the price will incur costs that do not add value.
- In strategic partnering the very virtue of the team rolling on from project to project can also be a disadvantage. The lack of new thinking, of a fresh pair of eyes, can mean that solutions are often sought in the narrow experience of previous projects. Value management offers the opportunity to redress this, both by brainstorming as a freer way of thinking and by introducing people from outside the immediate project team.
- There is also the danger of complacency. As the parties become more familiar with each other, in a deliberately non-adversarial atmosphere, performance levels may drop. In such an environment it becomes difficult to criticise the work of others, even where this falls below normally acceptable standards. This can lead to a loss of value both in the design and construction process and in the finished product.

9.3.3 Value management in a partnering environment in practice

For value management to make a contribution in a partnering arrangement, the value management process will need to be modified to take into account the differences between partnering and traditional procurement. The following issues should be considered:

- The value management workshop should introduce people from outside the immediate partnering team.
- More time may need to be spent on introducing the purpose and procedures of a brainstorming session, in order to encourage a genuine level of imaginative thinking.
- The workshop should capitalise on the well-established relationships, but should encourage the splitting up of established working subgroups. For example the engineers and architects will by necessity be used to working together; it may be helpful to split into smaller subgroups that throw together people less used to working together. This can help to generate more imaginative ideas.
- The introduction may need to spend less time on defining the project and more time on considering the lessons from earlier similar developments.
- More time can be spent on detailed functional analysis if it is intended is to roll out the improvements across a number of projects.

9.4 Value management, benchmarking and key performance indicators

In the previous section we highlighted some of the possible risks of partnering:

- *opportunistic behaviour* – if transactions are conducted without the element of competition
- *lack of innovation* – when the team rolls on from project to project
- *complacency* – as the parties become more familiar with each other, performance levels may be neglected.

9.4.1 Benchmarking

Benchmarking is one of the 'pillars' of partnering (see above), and provides a means of addressing these concerns. It provides not only a proxy for market price, without the need to go back to the market to check whether value is being obtained, but also awareness of performance in key areas, and the means to assess objectives of continuous improvement. This makes benchmarking an indispensable element of modern partnering.

9.4.2 Key performance indicators (KPIs)

An essential requirement for benchmarking is the ability to measure performance objectively across a set range of key targets. KPI is a generic term for any such important measure, but the UK Construction Best Practice Programme has published a standard list of 10 'headline' KPIs that have been generally adopted as representative of desirable performance:[7]

1. client satisfaction with the product
2. client satisfaction with the process
3. end-user satisfaction
4. time performance (in absolute terms)
5. cost performance (in absolute terms)
6. time predictability
7. cost predictability
8. safety
9. defects
10. profitability of the parties in the project.

All but two (safety and profitability) measure project performance, and are therefore either directly or indirectly measures of value.

9.5   Partnering and risk management

Partnering facilitates the application of risk management techniques and processes in much the same way as it does those of value management. Many exponents advocate combining the two into a single process, for example into a combined workshop.

The advantages that partnering brings to value management are equally applicable to the management of risks:

- opportunities to develop a long-term relationship
- well-integrated multidiscipline involvement, with a shared commitment to getting things right
- involvement of a contractor in the early stages of brief development and design
- involvement of the supply chain in the early stages
- opportunities for shared learning and retained team expertise
- the ability to access feedback from the building user or facility manager.

9.6   Private Finance Initiative: theory and practice

The Private Finance Initiative (PFI) is a procurement route for public facilities.
It is based on the procurement approach commonly referred to as
design–build–finance–operate (DBFO; see Chapter 2), and is a move away from
procuring of assets (in the form of buildings) and towards the purchase of services
(that the buildings provide).

The level of payment by the public sector is based on the performance of the
private sector operator against agreed levels of service. In theory this allows
the public sector improved value for money in partnership with the private
sector.

In the PFI the public sector client pays only on delivery of services to quality
standards that are usually specified by performance, rather than prescribed. The
private sector, normally in the form of a consortium (for an example see Chapter
2), makes its profit from providing some or all of the design, building, financing and
operation of the facility.

A project is more likely to be successful and acceptable as a PFI project if:

- the project has clear boundaries and measurable output performance
- there is scope for innovation in design, which enables the service provider
  to design away risks and bring new ideas to the way the service is
  provided
- the project has a substantial operating content
- there is scope for the service provider to find alternative uses for the asset
  provided
- any surplus assets intrinsic to the project are included in the package
- the risks transferred to the service provider are commercial in nature and
  controllable.[8]

9.6.1   The Private Finance Initiative and value management

The government's main objectives with the Private Finance Initiative and its later
version, public–private partnerships (PPP), are:

- to reduce the impact of large capital projects on the public sector borrowing
  requirement
- to seek projects that provide value for money
- to transfer construction and operating risks away from the public sector.

With regard to value, many of the advantages of partnering arrangements are also present in projects procured under the PFI:

- They provide the opportunity to work from the outset as an integrated team that includes contractors, designers, subcontractors and importantly those involved in the facilities management and funding of the project.
- Projects procured under the PFI allow, in theory, complete design freedom – there are no prescriptive building requirements, simply a series of output criteria to meet. How these are met can be the subject of an initial value management workshop that explores how best to achieve the service standard.
- There is focus on both cost and value. It is essential to focus both on the cost and on the quality of service; delivering a low-cost facility that does not provide both the quality standard and continuity of service provision will reduce the consortium's income stream. In extreme cases failure to deliver the required service quality may result in termination of the consortium's contract.
- There is likely to be a much greater focus on life cycle costs, as the running and maintenance costs of the building will be borne by the private consortium. There is an important secondary aspect to this: if there is any interruption to the supply of the service, for example to allow time for maintenance, the consortium will not receive payment. Therefore longevity and ease of maintenance will become important considerations.

Impacts of the PFI on value
According to the Treasury Task Force[9] the PFI has led to some significant improvements in value. For example, the average cost saving for the first eight DBFO road contracts was 15%, and the Bridgend and Fazakerly Prison project was estimated to represent a saving of more than 10% over contract life. These findings accord with figures revealed to the Construction Productivity Network.[10] From a survey of current PFI projects:

- cost savings were greater in transport and prisons than in accommodation, health and education
- in general, civil engineering projects showed greater savings than building projects.

Negative impacts upon value
There are also disadvantages to the PFI, which may have a negative impact on value:

- The process separates the public sector building promoter from the private sector consortium responsible for delivering and managing the facility. As the promoter is also likely to be the user, the effect is to separate the design team from those who will use the facility. In partnering, the intimate relationship between the user, the designers and the construction supply chain is a fundamentally sound arrangement (at least in theory) for delivering value. In the PFI the exact opposite arrangement exists, with all the implications that this has for value management.

- The split is not simply organisational but also contractual. Hence opportunities for value management workshops involving both promoters and the PFI team are in practice almost non-existent.

9.6.2 The Private Finance Initiative and risk management

One of the PFI's key aims is to effect a meaningful transfer of risk. In principle, risk should be allocated to whoever from the public or private sector is best able to manage it at least cost; it should not simply be dumped on the party least able to resist. The main reason for transferring risk is to create incentives for the private sector to supply cost-effective, high-quality services on time. The PFI consortium receives payments once the service begins to be provided. Payment continues only if the service meets the quality criteria set. In consequence risk reviews are fundamental to the PFI.

9.6.3 Partnering, the Private Finance Initiative and life cycle costing

In a partnering arrangement, and particularly in strategic partnering, there is a much greater commitment to the long term. The design and construction team will continue to be involved after the building is completed. Problems that arise in the maintenance or the energy consumption of the building will be shared with the designers and builders. This input will both help to build knowledge and a database and encourage proper consideration of these issues at the early project stages. The partnering team may well include people from a facilities management background who can contribute this knowledge at the outset.

Life cycle costing ought to be a primary concern of any private finance consortium. In practice, there is little evidence that life cycle costs are given any great consideration. Despite the incentive to reduce long-term running costs, and despite the proportion this represents of the building's whole life cost, the main emphasis is still on the capital cost of the facility. This is probably due to the same reasons that explain the generally low take-up of life cycle costing (see Chapter 7). As these barriers are broken down there is likely to be a much greater uptake of measures that reduce the life cycle cost of buildings procured under the Private Finance Initiative.

9.7   Conclusions

In this concluding chapter we have examined risk and value management from the point of view of projects based upon the ethos of partnering and those financed under the Private Finance Initiative. Both are important trends away from the traditional approach to project procurement. A brief introduction was given to the theory of each, and in each case there was an analysis of the role of value management and risk management. In both cases this appears to be an enhanced role.

The key points covered in this chapter were as follows:

- There are a number of advantages for value management with partnering.
- There are also some problems.
- The concept of benchmarking is critical to partnering, and answers these potential criticisms.
- Benchmarking depends upon measurement, for which key performance indicators are commonly used as measures of value.
- There are similar advantages for risk management with partnering as for value management.
- Risk and value are two key drivers of the Private Finance Initiative (PFI).
- Value (in terms of value for money) appears to be being achieved with the PFI.
- Life cycle costing is of fundamental importance to PFI projects.

Notes and references

1 Construction Task Force *Rethinking Construction: Report of the Construction Task Force on the Scope for Improving the Quality and Efficiency of UK construction (The Egan Report)* (DETR, London, 1998).

2 M. Latham *Constructing the Team: Joint Review of Procurement and Contractual Arrangements in the UK Construction Industry* (HMSO, London, 1994).

3 See, for example, M. Saad and M. Jones *Unlocking Specialist Potential* (Reading Construction Forum, Reading, 1998).

4 See http://www.local-regions.detr.gov.uk/pfi/3.htm#introduction (Department for Transport, Local Government and the Regions).

5 Source: *Best Value Through Partnerships* (www.4ps.co.uk).

6 J. Bennett and S. Jayes *The Seven Pillars of Partnering,* Reading Construction Forum (Thomas Telford, London, 1998).

7 Construction Best Practice Programme *Benchmarking Factsheet* (1998), available from http://www.cbpp.org.uk/cbpp/cbpp_publications/main.jsp

8 Building Design Easibrief.

9 Treasury Task Force *Partnerships for Prosperity: The Private Finance Initiative* (HM Treasury, London, 1998).

10 Construction Productivity Network *Value for Money as a Potential Driver for the PFI,* CIRIA members report 0148 (Construction Industry Research and Information Association, London, 2000). Presented at CPN meeting, 15 November 2000, London.

Index